MW00437572

ALCHEMY

E. J. HOLMYARD

DOVER PUBLICATIONS, INC.
NEW YORK

This Dover edition, first published in 1990, is an unabridged and unaltered republication of the work first published by Penguin Books, Harmondsworth, Middlesex, England, in 1957.

Manufactured in the United States of America
Dover Publications, Inc.
31 East 2nd Street
Mineola, N.Y. 11501

Library of Congress Cataloging-in-Publication Data

Holmyard, Eric John, 1891–1959.
Alchemy / E.J. Holmyard.
p. cm.
Reprint. Originally published: Harmondsworth, Middlesex, England : Penguin, 1957.
Includes bibliographical references (p.).
ISBN 0-486-26298-7
1. Alchemy. I. Title.
QD13.H64 1990
540′.1′12—dc20 90-3016
CIP

CONTENTS

LIST OF TEXT FIGURES

LIST OF PLATES

AUTHORITIES

As might be expected of a subject that had an active life of some two thousand years, the literature of alchemy is vast. Even the catalogue compiled by Mrs Charles (Dorothea Waley) Singer of western alchemical manuscripts runs into many volumes; oriental manuscripts are probably as numerous; and the printed alchemical books are legion. But alchemy has always interested historians of science and of the occult so that, though much remains unexamined, the main story of its rise, development, and degeneration can be followed.

The principal authorities, apart from original sources, that any student of alchemy must consult, and that have been used in the preparation of this book, are as follows:

Ambix, the Journal of the Society for the Study of Alchemy and Early Chemistry. This journal, which owes much to the scholarship of its founder editor, the late Dr F. Sherwood Taylor, himself an authority on Greek alchemy, is indispensable to any historian of alchemy.

Isis, a journal dealing with the history and philosophy of science, was founded and for many years edited by George Sarton (*ob.* 1956), whose monumental *Introduction to the History of Science* is a mine of accurate information on early scientists and their works.

Lynn Thorndike: *A History of Magic and Experimental Science from the Twelfth to the Sixteenth Century.* London and New York, 1923–41. This scholarly work has the uncommon merit of combining erudition with a pleasant and easy style, not seldom enlivened by flashes of dry humour.

John Ferguson: *Bibliotheca Chemica*, A Bibliography of Alchemical, Chemical, and Pharmaceutical Books. Glasgow, 1906.

E. O. von Lippmann: *Entstehung und Ausbreitung der Alchemie.* Berlin, 1919–31.

L. Figuier: *L'Alchemie et les Alchimistes.* Paris, 1856.

M. P. E. Berthelot: *La Chimie au Moyen Age.* Paris, 1893; and *Collection des Anciens Alchimistes Grecs.* Paris, 1888.

JOHN READ: *Prelude to Chemistry.* Second edition. London, 1939. Contains some information not to be found elsewhere.
F. SHERWOOD TAYLOR: *The Alchemists.* New York, 1949.
A. J. HOPKINS: *Alchemy, Child of Greek Philosophy.* New York, 1934.
E. ASHMOLE: *Theatrum Chemicum Britannicum.* London, 1652.
H. F. M. KOPP: *Geschichte der Chemie.* Braunschweig, 1843–7.

Many other authorities have been consulted on particular topics, but those listed above will give ample guidance to any reader who wishes to make a fuller study of alchemy than the present sketch of the subject can provide. The author expresses his grateful thanks to Mrs Singer for her kindness in reading the proofs. He alone is responsible for any errors that may remain.

E. J. H.

I

INTRODUCTORY

THE art or science of alchemy is of great antiquity, for it was practised before the birth of Christ. It has also had a long history, for there are still alchemists to be found, not merely in such less enlightened countries as Morocco and parts of the East, but in England, the United States, France, Italy, and Germany. Its heyday was from about A.D. 800 to the middle of the seventeenth century, and its practitioners ranged from kings, popes, and emperors to minor clergy, parish clerks, smiths, dyers, and tinkers. Even such accomplished men as Roger Bacon, St Thomas Aquinas, Sir Thomas Browne, John Evelyn, and Sir Isaac Newton were deeply interested in it, and Charles II had an alchemical laboratory built under the royal bedchamber with access by a private staircase. Other alchemical monarchs were Herakleios I of Byzantium, James IV of Scotland, and the Emperor Rudolf II. There are several references to alchemy in Shakespeare; Chaucer devoted one of his *Canterbury Tales* to it (p. 177); and Ben Jonson wrote a play, *The Alchemist*, in which he shows considerable knowledge of the subject. Romance and adventure, religious and mystical emotion, fraud and trickery, scientific inquiry, skilful technology, tragedy and comedy, poetry and humour, are all to be found on turning the variegated pages of its history.

Alchemy is of a twofold nature, an outward or exoteric and a hidden or esoteric. Exoteric alchemy is concerned with attempts to prepare a substance, the philosophers' stone, or simply the Stone, endowed with the power of transmuting the base metals lead, tin, copper, iron, and mercury into the precious metals gold and silver. The Stone was also sometimes known as the Elixir or Tincture, and was credited not only with the power of transmutation but with that of prolonging human life indefinitely. The belief that it could be obtained only by divine grace and favour led to the development of esoteric or mystical alchemy, and this gradually developed into a devotional system where the mundane transmutation of metals became merely symbolic of the transformation of sinful man into a perfect being through prayer and submission to the will of God. The two kinds of alchemy were often inextricably mixed; however, in some of the mystical treatises it is clear that the authors are not concerned with material substances but are employing the language of exoteric alchemy for the sole purpose of expressing theological, philosophical, or mystical beliefs and aspirations. In the present book we shall deal principally with exoteric alchemy, but this cannot be properly appreciated if the other aspect is not always borne in mind.

It has further to be remembered that the practical alchemists were well aware that if (they could not know that the emphasis was on the 'if') they succeeded in making gold artificially their lives might be in grave danger from the avaricious princes and other evilly disposed persons. Even the suspicion that they had discovered the secret was often sufficient to imperil them. One alchemist complained that, falling under this suspicion because he had happened to effect some rather spectacular cures during an epidemic, he had to disguise himself, shave off his beard, and put on a wig before he was able to escape, under a false name, from a mob howling for his elixir; he added that he knew of persons who had been found strangled in their beds simply because they were thought to have found the Stone, though in reality they knew no more about it than their murderers. It will appear in the following pages that the possession of alchemical lore was in fact a perilous liability, even when royal licences to practice the Art were granted, as they often were by Henry VI of England and other rulers.

For reasons of safety, therefore, as well as from a cupidity that did not wish to share knowledge that might prove invaluable, the alchemists used to describe their theories, materials, and operations in enigmatical language, efflorescent with allegory, metaphor, allusion, and analogy. Some of this language can be interpreted by one familiar with the literature and with the substances commonly used in alchemy, and no doubt more of it could be understood by the adepts themselves; but the result of such cryptic modes of expression is that it is not always possible to decide whether a particular passage refers to an actual practical experiment or is of purely esoteric significance. The point is referred to again in Chapters 2 and 7, but meanwhile it may be useful to provide a sample in illustration.

According to an anonymous seventeenth-century book entitled *The Sophic Hydrolith*, the philosophers' stone, or the ancient, secret, incomprehensible, heavenly, blessed, and triune universal stone of the sages, is made from a kind of mineral by grinding it to powder, resolving it into its three elements, and recombining these elements into a solid stone of the fusibility of wax. The details of the process are scarcely as simple as this outline would suggest. It is first necessary to purge the original material of all that is thick, nebulous, opaque, and dark in it, an operation to be effected by means of 'our Pontic water', which is sweet, beautiful, clear, limpid, and brighter than gold or diamonds or carbuncles. Then the extracted body, soul, and spirit must be distilled and condensed together by their own proper salt, yielding an aqueous liquid with a pleasant, penetrating smell, and very volatile. This liquid is known as mercurial water or water of the Sun. It should be divided into five portions, of which two are reserved while the other three are mixed together and added to one-twelfth their weight of the divinely endowed body of gold. Ordinary gold is useless in this connexion, having been defiled by daily use.

When the water and the gold have been combined in a solutory alembic (p. 48) they form a solid amalgam, which should be exposed to gentle heat for six or seven days. Meanwhile one of the two reserved fifths of the mercurial water is placed in an egg-shaped phial and the amalgam is added to it. Combination will slowly take place, and one will mingle with the other gently and imperceptibly as ice with warm water. This union the sages have compared to the union of a bride and bridegroom. When it is complete the remaining fifth of the water is added a little at a time, in seven instalments; the phial is then sealed, to prevent the product from evaporating or losing its odour, and maintained at hatching-temperature. The adept should now be on the alert for various changes. At the end of forty days the contents of the phial will be as black as charcoal: this stage is known as the raven's head. After seven days more, at a somewhat higher temperature, there appear granular bodies, like fishes' eyes, then a circle round the substance, which is first reddish, then white, green, and yellow, like a peacock's tail, a dazzling white, and finally a deep red. That marks the climax, for now, under the rarefying influence of the fire, soul and spirit combine with their body to form a permanent and indissoluble Essence, an occurrence that cannot be witnessed without admiration and awe. The revivified body is quickened, perfected, and glorified, and is of a most beautiful purple colour; its tincture has virtue to change, tinge, and cure every imperfect body.

That is, if everything has gone well; but sometimes mishaps threaten. There are four bad signs: a red oil floating on the surface, too rapid a transition from white to red, imperfect solidification, and refusal of a test portion of the substance to melt like wax when placed on hot iron. If these are not given immediate attention no success will be attained. If any of them should be observed, the compound must be taken out of the phial and treated with more of the mercurial water. It is then to be heated till any sublimation or evolution of vapour has ceased, when it may be replaced in the phial and the original treatment continued.

The author concludes by reminding the successful operator that the Stone thus prepared includes all temporal felicity, bodily health, and material fortune. By its aid Noah built the Ark, Moses the tabernacle with all its golden vessels, and Solomon the Temple, besides fashioning many precious ornaments and procuring for himself long life and boundless riches. Yet the Stone cannot be applied for purposes of metallic transmutation in the form in which it was left at the completion of the operation described, but must be further fermented and adapted; otherwise it could not be conveniently projected upon imperfect metals. The additional treatment consists in melting in a crucible one part of the Stone with three parts of the purest gold available, whereupon an efficacious tincture will be obtained capable of transmuting one thousand times its own weight of base metal into gold. Many other things may be done with the tincture which must not be revealed to this wicked world.

The word alchemy is derived from the Arabic name of the art, *alkimia,* in which 'al' is the definite article. On the origin of 'kimia' there are differences of opinion. Some hold that it is derived from *kmt* or *chem,* the ancient Egyptians' name for their country; this means 'the black land', and is a reference to the black alluvial soil bordering the Nile as opposed to the tawny-coloured desert sands. In the early days of alchemy it was much practised in Egypt, and if this derivation is accepted the name would mean 'the Egyptian art'. Against this etymology is the fact that in ancient texts *kmt* or *chem* is never associated with alchemy, and it is perhaps more likely that *kimia* comes from the Greek *chyma,* meaning to fuse or cast a metal. As practical alchemy dealt very largely with this particular operation, it might well have been named from it. Whatever the truth, our word alchemy and its modern formation, chemistry, come directly from the Arabic, and provide reminders that in the early Middle Ages the principal students of the Art were Muslims (Chapter 5).

The origins of alchemy itself were diverse. When men had become cultivators of the soil and stockbreeders, instead of mere food-gatherers, they took to building towns, thus inaugurating the change in methods of living known as the urban revolution. As a result of this revolution, communities were able to support specialized craftsmen on the surplus of the harvests procured by the agricultural workers, and by at latest 3000 B.C. such crafts as metallurgy, weaving, carpentry, building, and the making of dyes and pigments were well established. The art of writing and recording had also been invented, probably in Mesopotamia, one of the earliest known documents being a clay tablet of about 3600 B.C. giving a statement of the financial accounts of a temple.

During the 3000 years or so before the first definite appearance of alchemy in the last couple of centuries before Christ, the accumulation of technical knowledge went steadily on, and some of the achievements of ancient craftsmen have never been surpassed. Coloured alloys and artificial gems were manufactured, glass-making was well established, and the useful properties of very many minerals and plants had been discovered (Chapter 4). But all such familiarity with material objects and the changes that could be effected in them did not imply the segregation of what we should now call technology from the other aspects of daily life. The operations of the craftsmen were carried out to the accompaniment of religious or magical practices, and supposed connexions were seen between metals, minerals, plants, planets, the Sun and Moon, and gods. Thus in Babylonia gold was connected with the Sun and with the god Enlil, and silver with the Moon and the god Anu. Astrological considerations became of increasing importance, and by the sixth and fifth centuries B.C. a very complex science of astrology had been elaborated. Since many of the crafts later drawn upon by the alchemists, particularly metallurgy and colouring, were much influenced by the observance that had to be paid to astrological beliefs, it is worth while to examine this point more closely; with the operations they took over, the alchemists also accepted much of the astrological speculation.

In the first place, astrology emphasized a harmony between the macrocosm or universe and the microcosm or man; all that went on in the universe had its influence on, and its parallel in, man. The soul of man was believed to enter the body by way of a particular star, and at death to return to heaven by the same path. The signs of the zodiac, by then established as twelve, had a magical significance and could be used for casting horoscopes, not merely for man but for discovering the favourable conditions for carrying out, say, the preparation of a certain drug or alloy. The calculations involved in making the horoscope often required the use of mystic numbers such as magic squares, so that an esoteric numerology arose. Such a numerology was further developed by Pythagoras (*c.* 530 B.C.) and is frequently encountered in alchemical treatises (pp. 38, 76).

With the Greeks of about the fourth century B.C. astrology was still regarded as concerned with the regulation of all happenings in the universe, as it had been in ancient Mesopotamia, but whereas the Babylonian astrologers had given pride of place among the heavenly bodies to the Moon, the Greeks gave precedence to the Sun. The Moon and the five planets then known were assigned each to a special deity and endowed with the characteristics of that deity; on this system the reddish planet was called after Mars, the god of war, and astrologically governed warlike affairs, while the planet assigned to Venus was potent in matters of love. The old idea that the planets were connected with metals was also adopted, so that the Sun, the Moon, Mars, Mercury, Venus, Jupiter, and Saturn were often metaphorically used to signify gold, silver, iron, mercury or quicksilver ('argent vive'), copper, tin, and lead.

Besides astrology, other philosophical sciences were now being cultivated. Greek physicians and thinkers visited the centres of learning in Mesopotamia and Persia, and brought back ideas not merely from those centres themselves but from other visitors who had come from the opposite direction, namely from India, central Asia, and even China. All this crude material was worked up by such great philosophers as Plato and Aristotle into the imposing body of Greek thought that has fundamentally affected Western civilization ever since.

With their growing intellectual achievements the Greeks of this period combined military prowess, and under Alexander the Great (356–323 B.C.) they destroyed the Persian power, invaded north India, conquered Tyre and Gaza, and occupied Egypt. The last country, with its delightful climate and its air of inscrutable wisdom, attracted Greek settlers in great numbers, and in 332 B.C. Alexander founded in the Nile delta the city named after himself, Alexandria.

The stage was now set for the rise of alchemy, but before beginning our story proper it will be profitable to spend a little time in a brief examination of the views of Aristotle (384–322 B.C.) on the constitution of matter, for those views were to form much of the background of exoteric alchemical theory. According to Aristotle, then, the basis of the material world was a prime or primitive matter, which had, however, only a potential existence until impressed by 'form'. By form he did not mean shape only, but all that conferred upon a body its specific properties. In its simplest manifestation, form gave rise to the 'four elements', fire, air, water, and earth, which are distinguished from one another by their 'qualities'. The four primary qualities are the fluid (or moist), the dry, the hot, and the cold, and each element possesses two of them. Hot and cold, however, and fluid and dry, are contraries and cannot be coupled; hence the four possible combinations of them in pairs are:

> *Hot and dry*, assigned to *fire*.
> *Hot and fluid* (or *moist*), assigned to *air*.
> *Cold and fluid*, assigned to *water*.
> *Cold and dry*, assigned to *earth*.

This may be expressed diagrammatically as shown in figure 1.

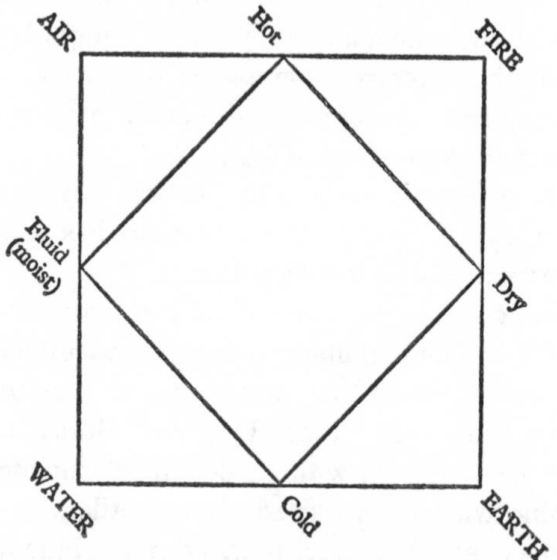

Figure 1. The Four Elements

In each element, one quality predominates over the other; in earth, dryness; in water, cold; in air, fluidity; and in fire, heat. None of the four elements is unchangeable; they may pass into one another through the medium of that quality which they possess in common; thus fire can become air through the medium of heat, air can become water through the medium of fluidity; and so on. Two elements taken together may become a third by removing one quality from each, subject to the limitation that this process must not leave two identical or contrary qualities; thus fire and water, by parting with the dry and cold qualities could give rise to earth. In all these changes it is only the 'form' that alters; the prime matter of which the elements are made never changes, however diverse and numerous the changes of form may be.

Aristotle next argues that each and every other substance is composed of each and every 'element', the difference between one substance and another depending on the proportions in which the elements are present. The kind of reasoning on which this proposition is based may be followed by observing what happens when a piece of green wood is heated; drops of water form at the cut end of the wood, therefore wood contains water; steam and vapours are given off, therefore wood contains air; the wood burns, therefore it contains fire; and an ash is left, therefore wood contains earth. A substance that contains much fire in its composition will burn easily; similarly a liquid owes its liquidity to the high proportion of water in it; a highly volatile substance contains much air; and a stone is preponderantly composed of earth.

Now, as we have seen, each element can be transformed into any of the others. It follows that any kind of substance can be transformed into any other kind by so treating it that the proportions of its elements are changed to accord with the proportions of the elements in the other substance. This may be done by change of the elements originally existing in the first substance, or by adding some substance consisting of such a proportion of the elements that when the two substances are mixed or combined the desired final proportions are attained. Here we have the germ of all theories of metallic transmutation and the basic philosophical justification of all the laborious days spent by alchemists over their furnaces. If lead and gold both consist of fire, air, water, and earth, why may not the dull and common metal have the proportions of its elements adjusted to those of the shining, precious one?

The question was a reasonable one, but left the more difficult one – as to the method employable – unanswered. Various practical methods suggested themselves to the alchemists in due course, as we shall see; and the idea of the harmony and unity of the universe, 'One is All, and All is One', led to the belief that the universal spirit could somehow be pressed into service either through the stars or by concentrating it, so to speak, in a particular piece of matter – the philosophers' stone. These were alchemical speculations, not Aristotle's, but they were founded on his cosmology.

Aristotle also expressed views on the formation of metals and minerals that helped to direct alchemical thought. He believed that there were two 'exhalations' concerned, but it is not very clear whether these are to be considered as material or spiritual; perhaps they were special modifications of the universal spirit. In any case, one of the exhalations is vaporous and the other smoky. The vaporous exhalation is formed when the Sun's rays fall upon water, and is moist and cold, while the smoky exhalation is formed when the rays fall upon dry land, and is hot and dry. Each exhalation is, however, mixed with more or less of the other. To the two exhalations correspond two classes of bodies that originate in the earth, namely minerals and metals. The heat of the dry exhalation is the cause of all minerals, or in other words these substances are composed mainly of the 'smoky' exhalation. Such are the kinds of minerals that cannot be melted, and realgar, orpiment, ochre, sulphur, and other substances of that sort. The 'vaporous' exhalation is the cause of all metals, those bodies that are either fusible or malleable or both, such as iron, copper, gold. All these originate from the imprisonment of the vaporous exhalation in the earth, the dryness of which compresses it and finally converts it into metal. Thus, since neither exhalation is entirely free from the other, metals and minerals, like all other substances, are composed of each of the four elements, but in metals the predominating elements are water and air (chiefly water), while in minerals they are earth and fire (chiefly earth).

To conclude this preamble, it may be recollected that the theory of the unity of the world permeated by a universal spirit had a corollary in the assumption that every object in the universe possessed some sort of life. Metals grew, as did minerals, and were even attributed sex. A fertilized seed of gold could develop into a nugget, the smoky exhalation was masculine and the vaporous one feminine, and mercury was a womb in which embryonic metals could be gestated. These and similar animistic beliefs mingle with the more rational outlook of Aristotle, and are more closely related to late forms of Platonism.

2
THE GREEK ALCHEMISTS

THERE is some doubt concerning the earliest mention of alchemy, for a reference to it occurs in a Chinese edict of 144 B.C. (p. 33), while a book on alchemical matters was written in Egypt by Bolos Democritos at a date that cannot be more precisely fixed than as about 200 B.C. However, whether the honour should go to China, or whether Egypt established a slight lead, there is no uncertainty about the fact that the main line of development of alchemy began in Hellenistic Egypt, and particularly in Alexandria and other towns of the Nile delta.

After its founding in 332 B.C., Alexandria rapidly grew to be the greatest and most important town of the ancient world. Under its rulers the Ptolemies an enormous library was gathered together – Ptolemy Philadelphus (285–247 B.C.) even being fortunate enough to acquire Aristotle's personal library – and a museum or university was built to house the scholars attracted thither from all parts of the Greek world. A mathematical school was started by the great Euclid himself, and among its celebrated pupils were Archimedes, whose 'Principle' is known to (even if not always understood by) every schoolboy; Hipparchus, who catalogued over a thousand stars; Eratosthenes, who measured the circumference of the Earth; and Apollonius of Perga, who wrote a treatise on conic sections. Grammar, literary criticism, philology, astronomy, astrology, and medicine all found learned teachers and enthusiastic disciples, and this intellectual activity was stimulated by foreign contacts arising from the port's thriving export and import trade.

Other towns in the Delta shared the delight in learning, and it was at one of them, Mendes, that Bolos Democritos wrote his book called *Physika*. This was divided into four parts, dealing respectively with the making of gold, the making of silver, the making of gems, and the making of purple. The recipes Bolos Democritos gives were collected from a variety of sources, such as craftsmen's notebooks and scraps of practical information from Egypt, Persia, Babylonia, and Syria, but what distinguishes him from the artisans is that he was interested in the transmutation of matter, which he thought was indicated by the changes in colour of metals when undergoing such treatment as alloying. He and other early alchemists therefore searched the collections of recipes made by metallurgists, dyers, glass-makers, and similar workers, to try to discover some processes that seemed likely to suit their purpose, often copying only parts of the recipes quite insufficient to carry the process through to the ends designed by the craftsmen. The object was to find some way of tingeing, tinting, dyeing, varnishing, or alloying one metal to make it resemble another – especially to make a base metal resemble gold.

From making a metal that resembled gold to believing that the artificial product really was true gold was only a short step for the alchemists, who lacked the technical training of the goldsmiths, and whose fundamental curiosity was philosophical rather than directed to mercenary gain. If a metal had a golden lustre, they thought, it must be gold, though Archimedes could have told them differently. And if artificial gold tarnished after a while it was merely because the transmutation had not been fully successful. Colour, in fact, to the alchemists was the most important characteristic of a metal, and so we find throughout Greek alchemical literature an insistence on colour-changes and sequences of colour-changes that left its mark on all subsequent alchemy. The underlying idea seems to have been that since the prime matter was the same in all substances, an approximation to this prime matter should be the first quest of alchemy; when such a substance had been obtained it was to be successively impressed with 'pure qualities which one after another should gradually rise in the scale of metallic virtue' to the perfection of gold. Manifold were the attempts to procure the basic matter, and any black solid made from non-precious metals by fusion, alone or with the addition of sulphur, arsenic sulphides, or other substances, was deemed a possibility. Gradually there grew up a generally accepted order in which the colours impressed upon this raw material should appear if the process were to be successful: black, white, iridescent, yellow, purple, red: but individual alchemists sometimes varied the order according to their own ideas or theories. An operation of this kind was often carried out in an apparatus known as a kerotakis (p. 49).

It cannot be doubted that such early alchemical practice and theorizing went on continuously from the time of Bolos Democritos, but unfortunately scarcely any records of it remain until we reach a period some 500 years later. Two papyri, known as the Leiden and Stockholm papyri, discovered in a grave at Thebes in Egypt, contain what seem at first sight purely practical recipes based on a book on dyeing written by Anaxilaos of Larissa in or about 26 B.C.; the papyri themselves probably date from about A.D. 300. Closer examination of the recipes, however, has shown that they could not yield any practical results, and it may therefore be that they were snippets collected by an alchemist. Stronger evidence that alchemy was being practised in the centuries immediately before and after the birth of Christ is provided by the fact that, about A.D. 300, an author named Zosimos, of Panopolis (Akhmim)) in Egypt, wrote an encyclopedia on the subject in twenty-eight books. Some of the passages in it are apparently original, but a large part of the work is a compilation from earlier texts now lost.

It is clear from the writings of Zosimos that, in the interval which had elapsed since Bolos Democritos wrote his *Physika*, alchemical speculation ran riot. We now find in it a bewildering confusion of Egyptian magic, Greek philosophy, Gnosticism, Neo-Platonism, Babylonian astrology, Christian theology, and pagan mythology, together with the enigmatical and allusive language that makes the interpretation of alchemical literature so difficult and so uncertain. Mercury, for instance, goes under many aliases: the silvery water, the ever-fugitive, the divine water, the masculine-feminine, the seed of the dragon, the bile of the dragon, divine dew, Scythian water, sea-water, water of the moon, and milk of a black cow. In order to give some show of authority to their nebulous doctrines, alchemists busied themselves in composing treatises that they then attributed to any philosopher or celebrity of earlier times whom their whim led them to select. Thus works on alchemy were ascribed to

Hermes, Plato, Moses, Miriam his sister, Theophrastus, Ostanes, Cleopatra, and Isis: in fact, if the support of any particular source was desired, steps were always taken to provide the appropriate books, sayings, or stories. Legends and myths were given alchemical interpretations; the golden fleece, which Jason and the Argonauts carried over the Pontic Sea to Colchis, was claimed to have been a manuscript on parchment, teaching the manner of making gold by alchemical art, and even the 'Song of Solomon' was supposed to be an alchemical treatise couched in veiled language.

Extant works of Zosimos were published with a French translation by Berthelot and Ruelle in 1887–8: they include his 'Authentic Memoirs', 'On the Evaporation of the Divine Water that fixes Mercury', and a 'Treatise on Instruments and Furnaces'. He tells us that the chemical arts were practised in Egypt under royal and priestly control, and that it was illegal to publish any work on the subject. Only Bolos Democritos had dared to infringe this regulation; as for the priests themselves, they had incised their secrets on the walls of the temples and pyramids in hieroglyphic characters, so that even if any seekers after forbidden knowledge were venturesome enough to brave the darkness of the sanctuaries they would have found the inscriptions unintelligible. The Jews, however, had been initiated into the mysteries and afterwards transmitted them to others.

Zosimos shows signs here and there of having had a fairly wide experience of chemical operations with metals and minerals. Thus he says that 'the second mercury', arsenic, can be obtained from sandarach (arsenic sulphide) by first roasting it to get rid of the sulphur, when 'cloud of arsenic' (arsenious oxide) will be left. If this is heated with various (reducing) substances it yields the 'second mercury', which can be used to convert copper into silver. The product of the last operation would of course not be silver, as he imagined, but copper arsenide, a white metal-like solid somewhat resembling silver in appearance. Zosimos also describes quite intelligibly the preparation of white lead from litharge and vinegar; the first product of this reaction, sugar of lead or lead acetate, is correctly said to be both sweet and salt-like, and, on keeping, to change slowly into white lead. Other chemicals mentioned are realgar, ochre, haematite, and natron, and Zosimos knew that mercury could be extracted from cinnabar.

Yet such genuine chemical knowledge forms a very insubstantial apex for the inverted pyramid of alchemical speculation erected upon it. Indeed, we may perhaps feel that the chemical facts were often introduced as a kind of seasoning, to give a flavour of authenticity to what were in reality pure products of the imagination. The apparent destruction of a metal by heat and its conversion to a powder, followed perhaps by regeneration of the metal on heating the powder with charcoal, could move Zosimos to write:

> I fell asleep and saw before me a priest standing upright before a bowl-shaped altar, which was approached by fifteen steps. The priest remained standing, and I heard a voice from on high which said to me, 'I have accomplished the action of descending the fifteen steps towards the darkness, and the action of ascending the steps towards the light. The sacrifice renews me, rejecting the dense nature of the body. Thus consecrated by necessity, I become a spirit.' Having heard the voice of him who stood upon the altar, I asked him who he was. In a feeble voice he answered me, 'I am Ion, priest of the sanctuary, and I have suffered intolerable violence. For one came quickly in the morning, cleaving me with a sword, and dismembering me systematically. He removed all the skin from my head with the sword that he held; he mixed my bones with my flesh and burned them with the fire of the treatment. It is thus, by the transformation of the body, that I have learned to become spirit.'

In later Greek alchemical writings, such symbolism and allegory become even more flowery. We may consider some of them here; others are quoted in another chapter (7). The best known of the Greek alchemists after Zosimos is Stephanos of Alexandria, who flourished at the time of the Byzantine Emperor Herakleios I (610–641). Herakleios, who overthrew the Persian might at the battle of Nineveh in 627, and restored the True Cross to Jerusalem, was not only a great soldier and statesman but an enlightened patron of learning. It is said that, like the later Holy Roman Emperor Rudolf II (p. 233), he was much interested in alchemy, the resemblance between the two monarchs in this respect extending to the fact that in their later years their absorption in alchemical studies led them to neglect affairs of state. Stephanos was in favour at the court of Herakleios at Byzantium, and was a philosopher, mathematician, astronomer, and alchemist. He gave public lectures on geometry, arithmetic, astronomy, music, and the philosophies of Plato and Aristotle, and wrote several books, including two on alchemy. The principal alchemical one is a long treatise divided into nine chapters or lectures; it gives a full exposition of the theory of alchemy as it was understood in the seventh century A.D. The Greek text has been studied in the seventh century with annotations, by Sherwood Taylor, who says that Stephanos was certainly not a practical laboratory worker but viewed alchemy as a mental process.

This opinion is based not only upon a general consideration of the work as a whole, but upon a particular passage, which Sherwood Taylor renders as follows:

> The ... wise man speaks in riddles as completely as possible. Thus he declares, as a teacher demonstrates everything, saying 'nothing is left remaining, nothing is lacking, except the vapour and the raising of the water.' Having shown in this the preparation of the whole, rendering all in a few words, that ye may not overwhelm the moving things with much matter, that ye may not think about saffron of Cilicia and the plant of anagallis, and the Pontic rhubarb for themselves, and of other juices, gall of quadrupeds and certain beasts, of stones and of destructive minerals, things that are dissimilar to the perfection-making, single and one nature, that men wandering shall not be led away from the truth, in order that in a natural existence they shall not seek for a non-existent tendency. What else? The most eminent man and counsellor of all virtue turns them around and draws them to the view of truth, that you may not, as I said [take note of] material furnaces and apparatus of glasses, alembics, various flasks, kerotakides, and sublimates. And those who are occupied with such things in vain, the burden of weariness is declared by them.

After this definite statement that practical alchemy is but a burden of weariness, it does not occasion any surprise to find Stephanos becoming lyrical about magnesia. The quotation is again from Sherwood Taylor's translation:

O wisdom of teaching such a preparation, displaying the work, O moon clad in white and vehemently shining abroad whiteness, let us learn what is the lunar radiance that we may not miss what is doubtful. For the same is the whitening snow, the brilliant eye of whiteness, the bridal procession-robe of the management of the process, the stainless chiton, the mind-constructed beauty of fair form, the whitest composition of the perfection, the coagulated milk of fulfilment, the Moon-froth of the sea of dawn, the magnesia of Lydia, the Italian stibnite, the pyrites of Achaea, that of Albania, the many-named matter of the good work, that which lulls the All to sleep, that which bears the One which is the All, that which fulfils the wondrous work. What is this emanation of the same Moon? I will not conceal it, but will display visibly the sought-for beauty. For the emanation of it is a mystery hidden in it, the most worthy pearl, the flame-bearing moonstone, the most gold besprinkled chiton, the food of the liquor of gold, the chrysocosmic spark, the victorious warrior, the royal covering, the veritable purple, the most worthy garland, the sulphur without fire, the ruler of the bodies, the entire yellow species, the hidden treasure, that which has the moon as couch ... For it is white as seen, but yellow as apprehended, the bridegroom to the allotted moon, the golden drop [falling] from it, the glorious emanation from it, the unchangeable embrace, the indelible orbit, the god-given work, the marvellous making of gold.

By the time of Stephanos, then, alchemy had very largely become a theme for rhetorical, poetical, and religious compositions, and the mere physical transmutation of base metals into gold was used as a symbol of man's regeneration and transformation to a nobler and more spiritual state. This convention was strengthened by Stephanos, and is characteristic of all late Greek alchemy; thus Archelaos, writing about 715, exclaims (C. A. Browne's translation):

When the spirit of darkness and of foul odour is rejected, so that no stench and no shadow of darkness appear, then the body is clothed with light and the soul and spirit rejoice because darkness has fled from the body. And the soul, calling to the body that has been filled with light, says: 'Awaken from Hades! Arise from the tomb and rouse thyself from darkness! For thou hast clothed thyself with spirituality and divinity, since the voice of the resurrection has sounded and the medicine of life has entered into thee.' For the spirit is again made glad in the body, as is also the soul, and runs with joyous haste to embrace it and does embrace it. Darkness no longer has dominion over the body, since it is a subject of light and they will not suffer separation again for eternity. And the soul rejoices in her home, because after the body had been hidden in darkness, she found it filled with light. And she united with it, since it had become divine towards her, and it is now her home. For it had put on the light of divinity and darkness had departed from it. And the body and the soul and the spirit were all united in love and had become one, in which unity the mystery had been concealed. In their being united together the mystery has been accomplished, its dwelling place sealed up and a monument erected full of light and divinity.

Though this imagery is drawn from amalgamations of metals with mercury, and from other alchemical operations, it is obviously in no sense to be regarded as a practical recipe: there are too many ambiguities and too few precise indications. The poem as a whole contains figurative references to many of the supposed stages of transmutation, but its significance is mystical or religious. The same may be said of other alchemical poems of about the same date, ascribed to Theophrastus, Hierotheos, and Heliodoros; they all follow the pattern of Stephanos and Archelaos. A line of cryptic signs, known as the 'formula of the crab' and supposed to be a hieroglyphic recipe for transmutation, is found in one of the works ascribed to Zosimos. Its concluding symbol is said to mean 'Blessed is he who understandeth', a pious ejaculation that might be uttered after reading any of the late Greek alchemists.

What the future development of alchemy in this direction might have been is happily needless to conjecture: events were now taking place that completely altered the entire scene. Before passing to those events, however, we must consider Chinese alchemy, the origin of which was approximately contemporaneous with that of the Alexandrian form of the art.

3

CHINESE ALCHEMY

NEW light upon the origins of alchemy has recently been provided by the discovery and publication of some early Chinese writings on the subject. It had long been known that the Chinese practised a mystical form of alchemy corresponding in many ways to that expounded by Western alchemical mystics, but the available data gave little information about the paths of development it had followed. At the present time, however, largely owing to the work of O. S. Johnson, T. L. Davis, L. C. Wu, K. F. Chen, and particularly H. H. Dubs, the facts are beginning to emerge – and very interesting facts they are. It appears, indeed, that one of the earliest historical mentions of alchemy is to be found in a Chinese imperial edict issued in 144 B.C., which enacted that coiners and those who made counterfeit gold should be punished by public execution. A commentator on this edict, writing about A.D. 180, explains that the Emperor Wen (about 175 B.C.) had allowed such practices, and much alchemistic gold had been made; however alchemistic gold is not really gold, and the alchemists thus lose their time and money and are left with nothing more than empty boasts. When they become poor through wasting their substance on their experiments, they turn to brigandry or robbery, and hence the Emperor Jing (Ching) issued his edict against them.

The fact that alchemy had to be prohibited by law indicates that it must have had a fairly lengthy previous history, and Chinese sources aver that it was first practised by a notability named Dzou (Tsou) Yen who flourished in the fourth century B.C. Dzou Yen was a wonder-worker highly esteemed by contemporary kings, nobles, and magnates, and professed to know what occult virtues should be studied in order that a state might prosper. It would have been in keeping if he had also promised his patrons that he would enrich them by alchemical means. Such an achievement would have been no more marvellous than his success in ripening millet in a district where the weather had been very cold. Dzou Yen played music on a set of warm pipes and the millet ripened at once.

In spite of the official ban on alchemy it was not eliminated, and in 133 B.C. – only eleven years after the edict – an alchemist was received by the Emperor Wu because he claimed that, by worshipping the goddess of the Stove, he had discovered the secret of immortality. The Emperor was interested, honoured the alchemist, and asked to be informed of the things to be done. The alchemist said that the Emperor must first worship the goddess of the Stove in his own person; this would enable him to invoke spiritual beings, who, when they appeared, would render possible the conversion of cinnabar into gold; this gold was to be made into vessels for drinking and eating, and even in itself would then prolong life sufficiently for the Emperor to give audience to 'the immortals of Penglai who live in the midst of the ocean'; and when this audience had been given with due sacrifices he would himself have become immortal.

This story is significant inasmuch as it reveals two characteristic features of Chinese alchemy, namely that the principal object of the art was to secure immortality or at least longevity, and that the help of spiritual beings or minor gods was to be invoked. Dubs tells us that the goddess of the Stove was 'a beautiful old woman clad in red garments with her hair done up in a knot on the top of her head. As the divinity in charge of cooking and brewing medicines, she naturally took care of alchemy too.' Commercial pursuits were despised in China at this period, which perhaps explains why accounts of attempts at the alchemical preparation of gold for mercenary purposes are few; the prolongation of life, on the other hand, was a noble aim, to which artificial gold should be solely devoted.

Experiments with this object were made for the Emperor Süan in 60 B.C. by a learned young man named Liu Hsiang (79–8 B.C.). The Emperor had become interested in the immortals and had had sacrifices made to the five sacred mountains and the four great rivers, a fact that the 'possessors of recipes' or magicians seized upon to induce him to establish official worship of a score of other divinities. Liu Hsiang, who in spite of his youth was a

Grandee Remonstrant or imperial adviser, improved upon the magicians by saying that gold could be made according to the recipe of Dzou Yen and others contained in two ancient books entitled 'The Great Treasure' and 'The Secrets of the Park'. The Emperor was sufficiently impressed to instruct Liu Hsiang to go ahead, at Treasury expense, and he spent a great deal of money before admitting failure. His lack of success gave officials the opportunity of impeaching him for having broken the edict of 144 B.C., and he was imprisoned and sentenced to death; but the Emperor valued his ability and easily allowed himself to be persuaded to cancel the sentence when Liu Hsiang's elder brother, the Marquis of Yangcheng, offered a substantial ransom.

The spectacular fiasco of imperially sponsored alchemy was ineffective in discouraging the optimism evinced by alchemists then and throughout the centuries, and explanations of Liu Hsiang's unsuccess were soon forthcoming. One is given in the following story of about the beginning of our era, related by Dubs. A gentleman of the imperial court, Cheng Wei, loved the art of alchemy. He took a wife, and secured a servant from a magician's household, after which he tried to make gold according to the recipes in 'The Great Treasure', but it would not come. His wife however stood by to watch him as he fanned the ashes to heat the retort containing quicksilver. She said, 'I want to try to show you something', and thereupon took a drug out of a bag and threw a very little of it into the retort. It was absorbed, and in a short while the contents of the retort had become silver. Wei was greatly astonished and asked his wife why she had not sooner told him that she possessed the secret. She replied 'In order to get it, it is necessary for one to have the proper fate.' In other words, a favourable destiny, recalling the favourable astrological conditions postulated by later alchemists, is a prerequisite; while the addition of the 'drug' foreshadows the indispensable 'philosophers' stone' of the future.

The next mention of alchemy is in the *Tsan-tung-chi* or 'Document Concerning the Three Similars' – possibly star-shaped hexagons, alchemy, and Taoist philosophy (p. 38) – ascribed to Wei Bo(Po)-yang. It was written about the second or third century A.D., which proves that its author used a pseudonym, for Wei Bo-yang or Bo-yang of Wei is a title of Lao-dze [Lao-tse] (*c.* 300 B.C. or perhaps a good deal earlier), founder of the Taoist (Daoist) system (p. 38). The book deals with the preparation of the 'pill of immortality', which is made from gold and which is so extremely efficient that it need be only very tiny. A Chinese biographical work entitled 'Complete Biographies of the Immortals' tells an amusing story of how Bo-yang tested the faith of some disciples in himself and his gold medicine (translated by Wu and Davis):

Bo-yang [and his white dog] entered the mountains to make efficacious medicines. With him were three disciples, two of whom he thought were lacking in complete faith. When the medicine was made, he tested them. He said, 'The gold medicine is made but it ought first to be tested on the dog. If no harm comes to the dog, we may then take it ourselves; but if the dog dies of it, we ought not to take it.' Bo-yang fed the medicine to the dog, and the dog died an instantaneous death. Whereupon he said, 'The medicine is not yet done. The dog has died of it. Doesn't this show that the divine light has not been attained? If we take it ourselves, I am afraid we shall go the same way as the dog. What is to be done?' The disciples asked, 'Would you take it yourself, Sir?' To this Bo-yang replied, 'I have abandoned the wordly route and forsaken my home to come here. I should be ashamed to return if I could not attain immortality. So, to live without taking the medicine would be just the same as to die of the medicine. I must take it.' With these final words he put the medicine into his mouth and died instantly.

On seeing this, one of the disciples said, 'Our teacher was no common person. He took the medicine and died of it. He must have done that with special intention.' The disciple also took the medicine and died. Then the other two disciples said to one another, 'The purpose of making medicine is to attempt at attaining longevity. Now the taking of this medicine has caused deaths. It would be better not to take the medicine and so be able to live a few decades longer.' They left the mountain together without taking the medicine, intending to get burial supplies for their teacher and their fellow disciple.

After the departure of the two disciples, Bo-yang revived. He placed some of the well-concocted medicine in the mouth of the disciple and in the mouth of the dog. In a few moments they both revived. He took the disciple, whose name was Yü, and the dog, and went the way of the immortals. By a wood-cutter whom they met, he sent a letter of thanks to the two disciples. The two disciples were filled with regret when they read the letter.

A good deal of information about Chinese alchemy is to be found in a treatise written by Go(Ko)-Hung, who lived in south China from 254 to 334, and used the sobriquet of Bao-pu-dzu (Pao P'u-tsu). This name literally translated means 'The master who preserves his pristine simplicity', but was happily rendered by Wu and Davis as 'Old Sober-sides'. The first part of Go-Hung's book is entitled *Nuy pe-en*, or 'Inner Chapters', and deals with the transmutation of metals, elixirs of life, ascetic rules for reaching longevity, and methods of achieving immortality. According to Dubs, Go-Hung maintains that other things besides chemical operations are necessary for success.

The alchemist must previously fast for a hundred days and purify himself by perfume. Officials, such as Liu Hsiang, cannot perform the necessary fasting. Only two or three persons should be present at a transmutation. It must be done on a famous great mountain, for even a small mountain is inadequate. It is impossible to perform a transmutation in a palace. The adept must moreover learn the method directly from those skilled in the art. Books are inadequate. What is written in books is only enough for beginners. The rest is kept secret and is given only in oral teaching. Worship of the proper gods is necessary. The art can moreover only be learned by those who are specially blessed. People are born under suitable or unsuitable stars. Above all, belief is necessary. Disbelief brings failure.

Go-Hung gives very numerous recipes for transmutations and for preparing elixirs, and points out that alchemical gold, which is uniform throughout, should be distinguished from a base metal superficially tinted to make it look like gold. He states that a man may prolong his life by taking medicines made from plants, but can become immortal only by the use of a Divine Elixir made from metals and minerals. It is difficult to identify the substances that were to be employed in the preparation of this elixir, but red and yellow arsenic sulphides, sulphur, cinnabar, alum, salt, white arsenic, oyster shells, mica, chalk, and the resin of the pine tree were certainly included among them. The resulting elixir, when thrown on to mercury or a mixture of lead and tin, converted the metal into gold or silver – gold from mercury, silver from the lead-tin alloy. Taken as a medicine for one hundred days it made a man immortal. An interesting detail is that Go-Hung gives what is probably the earliest description of the preparation of mosaic gold (stannic sulphide, yellow sulphide of tin).

The background of Chinese alchemical theory was the philosophy or religion of Taoism, usually said to date from *c.* 300 B.C. *Tao* (Dao) means the Way or Path of the Universe, and Taoism rests on the belief that the First Cause is the revolution of the heavens round the Earth. Soothill, quoting a writer of the fourth century B.C., says, 'What there was before the Universe was *Tao*: *Tao* makes things what they are, but is not itself a thing. Nothing can produce *Tao*: yet everything has Tao within it, and continues to produce it without end.' The Taoists wished for long life, the better to prepare themselves for Paradise, and to this end practised meditation, control of breathing, and various physical exercises, as well as frugality of diet. By a natural evolution, the desire for longevity grew into the hope of immortality, and the disciples of the cult were thus led to the study and practice of alchemy.

Passing from the general to the particular, we find three main points of relevance. The first is the great importance attached by the Chinese to the number 5; the second concerns the relation of this number to the magic square of the numbers 1 to 19 and the third is the theory of *Yin* and *Yang*. We may consider these three points in turn.

The significance given to the number 5 is reflected in the facts that there were five elements, namely wood, fire, earth, metal, and water; five zones of space; five 'directions', namely, north, south, east, west, and centre; five colours, namely yellow, blue, red, white, and black; and five stones from which man was first taught to extract copper. In Chinese alchemy the five elements, directions, and colours were related to one another and to the five metals gold, silver, lead, copper, and iron; thus earth was connected with the colour yellow, the direction centre, and the metal gold. The other groups were wood, blue, east, lead; fire, red, south, copper; metal, white, west, silver; and water, black, north, iron. A further connexion was between these groups and the five planets, water corresponding to Mercury, fire to Mars, wood to Jupiter, metal to Venus, and earth to Saturn. These equations help to elucidate many of the ideas and theories found in Chinese alchemy, and foreshadow similar supposed correspondences in later times.

The relation between the number 5 and the magic square of the first nine digits (excluding zero) (p. 76), has been examined by Stapleton, who points out that this square forms the ground plan of the *Ming-Tang* or imperial 'Hall of Distinction' or 'Temple of Enlightenment'. This was a square temple with nine rooms numbered according to the arrangement of the digits in the magic square, where the principal number, five, occupies the centre. The *Ming-Tang* 'seems to have been chiefly used for the promulgation of the monthly ordinances, and especially for the proclamation of the Calendar regulations, necessitated by the Chinese year having a variable length. Apart from the central room, each of the four rooms bearing an odd number had a single canopied Dais, whereas each of the even-numbered corner rooms had two such raised platforms. This gave a total of 12 such sites for the necessary monthly Proclamation of Space and Time.'

Stapleton suggests that, apart from the central room 5, the rooms may not originally have been deliberately numbered according to the magic square; but that, when by accident this arrangement was hit upon, a mathematician happened to notice its peculiar features. The *Ming-Tang* in that case would have been the origin of the magic square; but, whatever the truth, the plan of the temple became of importance in alchemy. This was because the Son of Heaven, when in the *Ming-Tang*, was believed to become the incarnation of the Deity and therefore possessed of unlimited power over matter. By using the ground-plan of the temple as an alchemical amulet or talisman, the adept would receive some of this power at second hand and would thus be able to prepare the pill of immortality or an elixir for transmuting base metals into gold. The chemical operations involved in these two processes required that the heating of the mixtures should be carried out five, or a multiple of five, times.

Some centuries after the appearance of the *Ming-Tang*, a new conception arose in Chinese philosophy, namely that of the 'Opposing Principles' *Yin* and *Yang*. It probably dates from about the sixth century B.C. The central idea of the theory was that the prime matter of the universe gave rise to two principles with directly opposed natures; thus *Yin* is feminine, watery, heavy, passive, and earthy, while *Yang* is masculine, fiery, light, and active. *Yang* was associated with the Sun and *Yin* with the Moon. Interaction between the two principles led to the formation of the five elements of which the world is constituted. Alchemical extension of this idea led to a connexion of *Yang* with gold, sulphur, cinnabar, and other substances supposed to have the power of giving life and fullness of years, though it does not appear that *Yin* was associated with mercury. The fact that *Yang* was regarded as the male fertilizer and *Yin* as the passively receptive feminine principle led to an emphasis on *Yang* substances in recipes for preparing the pill of immortality and the elixirs: but the practical instructions do not lead us very far along the Path, as the following example, translated by Wu and Davis from the *Tsan-tung-chi*, will illustrate:

Cooking and distillation takes place in the cauldron; below, blazes the roaring flame. Afore goes the White Tiger leading the way; following comes the Grey Dragon. The fluttering *Chu-niao* [Scarlet Bird] flies the five colours. Encountering ensnaring nets, it is helplessly and immovably pressed down, and cries with pathos like a child after its mother. Willy-nilly it is put into the cauldron of hot fluid to the detriment of its feathers. Before half the time has passed, Dragons appear with rapidity and in great number. The five dazzling colours change incessantly. Turbulently boils the fluid in the *ting* [furnace]. One after another they appear to form an array as irregular as dog's teeth. Stalagmites which are like midwinter icicles are spat out horizontally and vertically. Rocky heights of no apparent regularity make their appearance, supporting one another. When *yin* [negativeness] and *yang* [positiveness] are properly matched, tranquility prevails.

In other words, 'the solution was evaporated to crystallization'. How much more picturesque our textbooks of practical chemistry might be written!

The question of the relation between Chinese alchemy and that of Islam and the West is one that has given rise to much discussion. Transmission of knowledge and ideas was much more widespread in the ancient world than has sometimes been supposed, so that we might expect to find Chinese alchemical information affecting the practitioners of Persia, Mesopotamia, Arabia, and Egypt, as well as itself being influenced by a stream in the other direction. According to some scholars, the early Alexandrian alchemists were not attempting to make gold artificially, but were merely trying to prepare passable imitations. In support of this thesis they adduce the facts that there are no undoubtedly genuine references to full metallic transmutation much earlier than Islamic days, and that the Egyptians and Greeks knew far too much about metallurgy and assaying to render the alchemists' claims for their products to be credited. In China, on the other hand, definite attempts at transmutation had been made as early as the first century B.C. (p. 35), and at the same time the rarity of gold in that country meant that much less was known about its chemical and physical properties; alchemical gold was thus more likely to be accepted as authentic. They also point to the quite considerable trade going on between China and Alexandria in the first few centuries of our era, and altogether make out a good case for supposing that the idea of transmutation was brought by Chinese travellers.

There are, however, equally strong arguments on the contrary side. In the first place, Chinese alchemy was very predominantly intent upon preparing a medicine that should ensure immortality: but this alchemical trend does not appear in Near Eastern alchemy until the days of Islam, after the first Arab ships had dropped anchor at Canton in 714 and so initiated what became a very brisk trade. It seems hardly likely that if the Chinese brought stories of metallic metamorphoses to Alexandria they would not have accompanied them with stories of the pill of immortality. Further, expert metallurgical knowledge in the eastern Mediterranean countries would not necessarily lead to a disbelief in the possibility of transmutation, for not every specimen of artificial gold was submitted (or designed to be submitted) to assay. Among early workers on metals and minerals the belief that one metal might be changed into another could well have arisen independently at many different times and in many different countries, and the emphasis of Chinese alchemy on immortality and that of the Near Eastern on the acquisition of gold seem to imply that there was no close or genetic relationship between the two. The most that can be said with certainty is that there was exchange of information between China and Alexandria and, later, Islam; to settle the larger question of affinity and lineal descent many more data would be necessary than are at present available.

At the period when alchemy was beginning to grow rapidly in the Levant, it had already begun to dwindle in China. By A.D. 1000, experimental alchemy was virtually abandoned, and the vocabulary and terminology of the art had been adapted to spiritual and mystical systems – a development that had already begun in the West also. The quest for immortality was raised to a higher plane, but the scaffolding of alchemical thought was too useful to be abandoned.

4
ALCHEMICAL APPARATUS

BY the time that alchemy arose, various branches of technology had already reached a high degree of efficiency. Among them were metallurgy, ceramics, glass-making, dyeing and colouring, weaving, brewing, and the preparation of drugs, poisons, and cosmetics. The metallurgists knew how to extract such metals as copper and iron from their ores, and were thoroughly familiar with methods of assaying gold and silver with fair accuracy. Cupellation was one of the methods employed. Here the impure precious metal is mixed with lead and the metals are fused together in a porous crucible or cupel, often made of bone-ash. On blowing air over the melted mass, the lead and other base metals are oxidized, and the molten lead oxide or litharge, which contains all the base-metal oxides, is partly blown off by the blast, and partly absorbed by the walls of the cupel. Left in the crucible is a button of refined gold, or, if silver was originally present as well as gold, a button of gold-silver alloy. It was known as early as the second century B.C. that silver could be separated from gold by a modification of this cupellation method, namely by adding salt and barley husks as well as lead to the crude metal in the cupel. On heating, the silver is converted into silver chloride, which is absorbed by the cupel, and the gold is left in a state of purity.

Similar proficiency was shown by the dyers. Even as long ago as 1500 B.C. the Egyptians could produce fabrics dyed in stripes of red, blue, and yellow, and the simple dyeing of linen and leather had been practised for centuries earlier. The celebrated Tyrian or 'Imperial' purple dye was extracted on an industrial scale from the molluscs *Purpura* and *Murex* on the coast of the eastern Mediterranean between Tyre and Haifa. The streets of Tyre were notorious for their evil smell, which emanated from the decomposing remains of millions of these shellfish, only a small part of the body being used to obtain the dye. The art of mordanting with alum and other substances was also well known, as was the somewhat difficult process of dyeing with indigo. The dyer's was a skilled and respected craft – but apparently not a very pleasant one, for a papyrus of about 2000 B.C. says of an indigo-dyer, 'His hands stink, they have the odour of rotten fish, and he abhorreth the sight of all cloth.'

The making of cosmetics occupied a large number of workers, particularly in ancient Egypt, where black and green eye-paints were made from soot, burnt almond shells, or a black mineral such as stibnite or galena, and from the green mineral malachite and a green resin from certain conifers. Lip-colour was made from a basis of animal fats, and perfumes were extracted from flowers and aromatic resins. For the latter purpose, flower petals were dipped into fat or oil at about 65° C., or laid on a layer of solid animal fat, and replaced from time to time by fresh petals until the fat became fully charged with the perfume. Pomades obtained in this way were made up into balls or cones and were often worn on the head. In some cases a kind of distillation was used for extracting perfume from flower petals or leaves, and this process is of very great antiquity; a primitive form of distillation-apparatus dating from about 3500 B.C. has been unearthed at Tepe Gawra in north-east Mesopotamia and described by Martin Levey of Pennsylvania State University. It consists of a double rimmed pot, with holes in the inner rim to drain the channel between the two. The substance from which the perfume was to be extracted was placed in the channel, and a volatile oil or other liquid poured into the pot, over which a lid was then placed. On heating, the liquid boiled and the vapour condensed on the cooler lid. The liquid so formed ran into the channel, dissolved the perfume, and then flowed back into the pot, where it was volatilized again; hence the process was continuous, and was in fact what we now describe as a reflux distillation.

The manufacture of glass had been carried out in Egypt and Mesopotamia since late prehistoric times, the Assyrians even being expert enough to produce a wide range of coloured glasses, including a purple one containing colloidal gold. Glass jars and phials were common in Egypt, some of them decorated with glass threads of various colours. The Egyptians also made artificial pearls of glass beads and exported them to neighbouring countries. Early glass vessels were moulded, or built up from glass rods, but in the first century B.C. or thereabouts the art of glass-blowing was invented, in Syria. This very greatly increased the repertoire of the art, and glass vessels of all shapes and sizes soon became available. The glass-ware of Askalon in Syria long maintained a pre-eminent position.

From the preceding paragraphs it will be apparent that the first alchemists found ready to hand a mass of technological knowledge and practice, much of it very appropriate to the operations they wished to carry out. At the same time they were innovators in the sense that they often considerably modified existing apparatus and appliances, and adapted industrial chemical processes to ends for which they were not originally designed. The main alchemical operations were calcination, sublimation, fusion, crystallization, and distillation; they remained more or less constant throughout the alchemical period, though the apparatus employed underwent gradual improvement and elaboration, and with the discovery of powerful solvents such as alcohol, the alkalis, and the mineral acids a shift of emphasis from more drastic methods to milder ones was not uncommon. Those alchemists who used animal and vegetable products in their work frequently relied upon a further operation, namely fermentation or putrefaction, and this term was sometimes applied to changes in inorganic substances taking place with the evolution of gases. Gases themselves were not recognized as chemical individuals but were regarded as contaminated air or merely bad smells.

Calcination implied the reduction of a solid to the state of fine powder, principally by means of heat and generally with a change of composition. Thus although gold could be obtained as a powder, and was then known as calx of gold, it was in fact still in the metallic state; but when lead is heated in air it is converted into a yellowish-brown powder of a non-metallic nature – in this case the calx of lead is lead oxide. For calcining by means of heat many different kinds of furnaces were used, as they were for most of the other operations; hence large furnaces, small furnaces, and furnaces of intermediate size formed the major part of the equipment of an alchemical laboratory. The principal fuel was charcoal, though almost anything that would burn was pressed into service – wood, peat, rushes, oil, wax, pitch, coal, and dried dung of horses, cattle, and camels. It was a common obsession with the alchemists that if they could get a sufficiently high temperature, transmutation would be easy. They therefore used bellows to such an extent as to earn the nickname of *souffleurs*, 'puffers', bestowed upon them by the cynical. A glance at a picture of an alchemist's workshop (plates 1, 2, 3) will plainly reveal the supreme position held by the furnace among the multifarious tools and appliances that usually encumbered every square foot of space. The multiplicity of furnaces was rendered necessary not only by the need for economy of fuel but by the difficulty of regulating a single furnace to give different degrees of heat, a difficulty not overcome until the invention of dampers, said to be due to the fifteenth-century Bristol alchemist Thomas Norton (p. 188). Norton claimed that he had constructed a furnace in which no fewer than sixty different temperatures could be attained at one and the same time – but drawings of his laboratory nevertheless show several furnaces. Further regulation of furnace-temperature by the use of a chimney to promote the flow of air over the fuel was introduced by another alchemist, Johann Glauber (1604–68), in the seventeenth century. Earlier chimneys served merely to carry off fumes.

Fusion was usually effected in earthenware crucibles, either single or in the form of *bot bar bot*. This consisted of a lower crucible, over which was set a second with a perforated base. Crude metal or ore was mixed with a flux in the upper part of the apparatus, and on heating in the furnace the molten metal obtained flowed down into the lower crucible, leaving the slag or scories behind. Sublimation (p. 56), which consists in heating a substance until it vaporizes and then condensing the vapour directly back to the solid state by rapid cooling, is not of general application since most vapours pass through a liquid phase before solidifying. However, many of the substances used in alchemy happen to produce sublimates very easily; they include sulphur and many sulphides, amber and other resins, and camphor. The word sublimate was sometimes used to include a liquid product if it stuck to the top of the subliming apparatus, as is implied in a definition given by Geber (p. 134): 'Sublimation is the elevation of a dry thing by fire, with adherency to its vessel', but strictly speaking a sublimate should be solid.

Instructions for the building of furnaces often form part of alchemical treatises. Thus Geber says that the calcinatory furnace should 'be made square, in length four feet, and three feet in breadth, and let the thickness of the walls be half a foot; after this manner: Luna (silver), Venus (copper), Mars (iron), or other things to be calcined, must be put into dishes or pans of most strong clay, such as of which crucibles are made, that they may persist in the asperity of fire, even to the total combustion of the thing to be calcined.' He enthusiastically adds, 'Calcination is the treasure of a thing; be not you weary of calcination.' The athanor – from the Arabic *al-tannur*, furnace – resembled the calcinatory furnace, but contained a deep pan full of sifted ashes. The vessel with the matter to be heated was to be firmly sealed and then placed in the midst of the ashes, so that the thickness of the ashes surrounding the vessel, underneath and above, should be about four inches on the average, but more or less according to the intensity of the heating required.

According to this author, therefore, the athanor was a rather elaborate form of the heating-apparatus known to a modern chemist as a sand-bath. On the other hand, the solutory or dissolving furnace was a water-bath, consisting of a small furnace supporting a pan full of water with iron grids or rings in it on which glass vessels could be set. In the descensory furnace, an iron or earthenware funnel with a lid was fixed in the middle of the fire, and any liquid products flowed down the stem of the funnel into a receiver placed below.

If calcination and similar operations gave rise to much diversity of furnaces, distillation was responsible for a parallel array of stills. Some early forms are shown in Greek alchemical manuscripts of the first centuries of our era, which illustrate flasks, receivers, and other pieces of apparatus (plate 11). A typical still of this period consisted of a flask (*bikos*, cucurbite) to contain the liquid to be distilled, with a still-head carrying a delivery-spout or *solen.* The still-head was known as an *ambix,* but this name was often applied later to the still and still-head as a whole, and has given rise, by way of the Arabic *al-anbiq,* to our word alembic. Some still-heads were provided with two or three spouts, and the apparatus was then described as a *dibikos* or *tribikos.* The separate spouts do not appear to have been designed to collect different fractions of the distillate, so that their purpose is obscure. The invention of the *tribikos* was ascribed

Fig. 2. A Still, showing flask or cucurbit, alembic or still-head, delivery spout, receiver, and furnace

to a woman alchemist, Mary the Jewess, traditionally supposed to be Miriam, sister of Moses. In a work passing under her name, quoted by Zosimos (p. 27), instructions for making a *tribikos* are given. Three copper or bronze tubes should be made, each a cubit and a half in length, and of a thickness of metal rather greater than that of a pastry-cook's frying-pan. One end of each tube should be adjusted to the size of the neck of its receiver, and the other should be soldered into a copper or bronze still-head. The still-head should fit closely on to the earthenware vessel containing the matter to be distilled, and the joint should be luted with flour-paste. Similar luting was to be applied between the delivery-spouts and the receivers.

In later times, a special luting known as 'philosophers' clay' or 'clay of wisdom' was highly recommended. According to one recipe, it consisted of two-thirds clay free of stones and one-third of a mixture of dried dung and chopped hair.

Mary the Jewess is also credited with the invention of the water-bath – which is still known in France as the *bain-marie* – and of the apparatus called the kerotakis that figures prominently in Greek alchemical writings. In those days painting was frequently executed by the encaustic process, in which the pigments were mixed in melted wax – generally beeswax – and applied to the support by means of a brush. The metallic plate or palette, often triangular in shape, on which artists kept their paints fluid by resting it over a small charcoal-burner, was the kerotakis. In its application to alchemical operations, the kerotakis was placed inside a cylindrical or spherical container, closed at the lower end, containing mercury or sulphur or some other substance that would partly or entirely vaporize on heating. On the kerotakis proper (the name was later applied to denote the apparatus as a whole) was placed the metal to be subjected to the action of the vapours, normally in the form of foil or powder, and the orifice at the top of the cylinder or sphere was closed with a hemispherical cover. On heating, the volatile substance gave off its vapours, which in part attacked the metal and in part were condensed at the top of the apparatus, the liquid flowing back to the bottom. In effect, therefore, a continuous reflux action was maintained. The nature of the product varied according to the metals and vapours used; thus lead and copper on the kerotakis and sulphur below yielded a black substance that received much attention from the alchemists. Blackness or melanosis was supposed to represent the first stage towards transmutation; theoretically it should be followed by a whitening or leukosis and then by a yellowing or xanthosis, with sometimes a further stage of iosis or purpleness. Reconstructions of two forms of kerotakis by F. Sherwood Taylor are shown in figure 3.

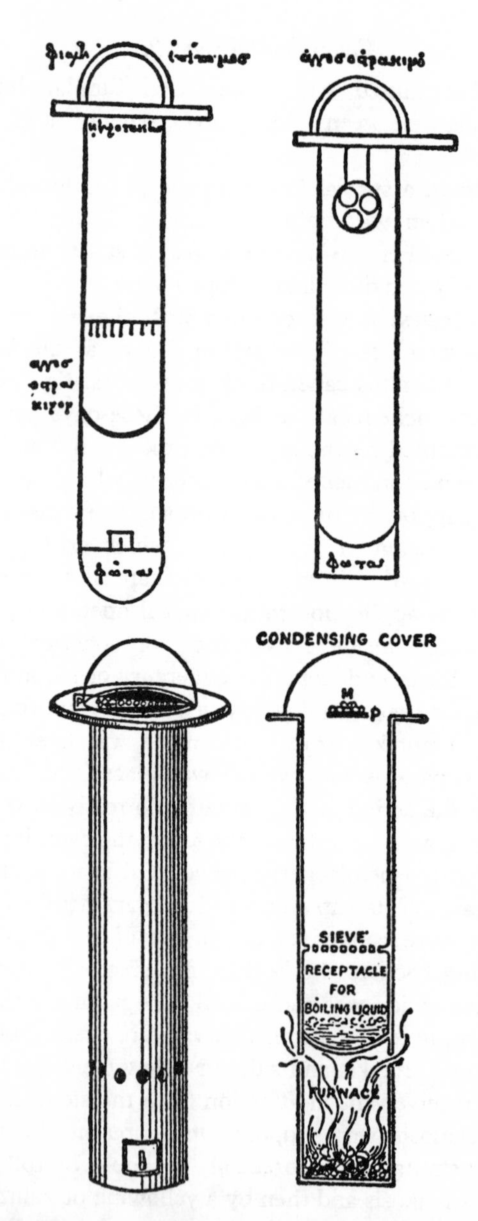

Figure 3. Kerotakides. Above, as shown in a Greek manuscript; below, conjectural restorations by F. Sherwood Taylor. (By courtesy of the *Journal of Hellenic Studies* and the executors of the late Dr Sherwood Taylor)

In early stills there was no efficient cooling (figure 4), so that volatile liquids were usually lost, though some could be absorbed in wool or cloth and afterwards squeezed out. There were many variations and combinations of alembics, but fundamentally

Figure 4. Distillation. Note the astrological influences in the top right-hand corner, and the optimism with which the receiver for the distillation is set. (From E. Darmstaedter, *Die Alchemie des Geber*, by courtesy of Springer-Verlag, Berlin)

they were all the same; the best they could do with moderately volatile substances is exemplified by their use in extracting essential oils from flowers. Here the use of batteries of alembics and receivers was applied on a commercial scale in the preparation of rose-water, of which the Caliph Al-Ma'mun (813–33) received an annual tribute of 30,000 phials from the neighbourhood of Shiraz. Fresh rose-petals were distilled gently, and the distillate consisted of an aqueous liquid containing the rose-oil. Weaker rose-water could be obtained by suitably diluting the distillate with water which had been purified by continued boiling and allowed to cool with protection from dust. For alchemical purposes, many other aromatic flowers, fruits, and leaves were treated in the same way, and the decoctions were used in attempts to prepare the elixir.

A curious belief commonly held by alchemists was that, even if they had discovered a process that would lead to success, the probabilities were that the substances used would have to be repeatedly subjected to this process – perhaps for hundreds of times – before the glorious end of the work revealed itself. It was a very ancient and persistent belief, for we find it among the Chinese alchemists (p. 39) no less than among later adepts such as Thomas Charnock (p. 199), who repeated one operation

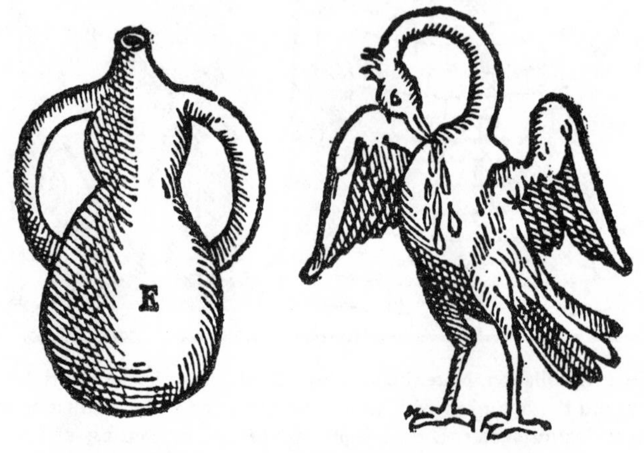

Figure 5. Pelican: *left*, alchemical; *right*, ornithological. (From J. B. Porta, *De Distillatione*, lib. ix, Rome, 1608)

476 times and hoped to continue to the five hundredth. The process of returning a distillate to its residue and then redistilling was called cohobation, and a special sort of double-reflux still could be employed to render the process automatic. This still was called the pelican, for reasons that will be obvious from the drawing (figure 5). It is thus very fitting that a book on alchemy should appear under the Pelican imprint.

The introduction of stills in which the condensation of vapours was effected outside the still-head appears to have been a European improvement of the twelfth or thirteenth century; at least, such stills are not mentioned by Islamic alchemists until after 1300, by which time they were well known in Europe. One of the first results was the preparation of alcohol in a state of comparative purity. It is not known with certainty who discovered alcoholic distillation, but a good case for the distinction might be argued in favour of Salernus, a physician of Salerno who died in 1167. About a century later, Taddeo Alderotti (1223–95), of Florence, describes the concentration of alcohol by repeated fractional distillation from a tightly luted alembic, the still-head of which was fitted with a delivery-spout a yard long connected

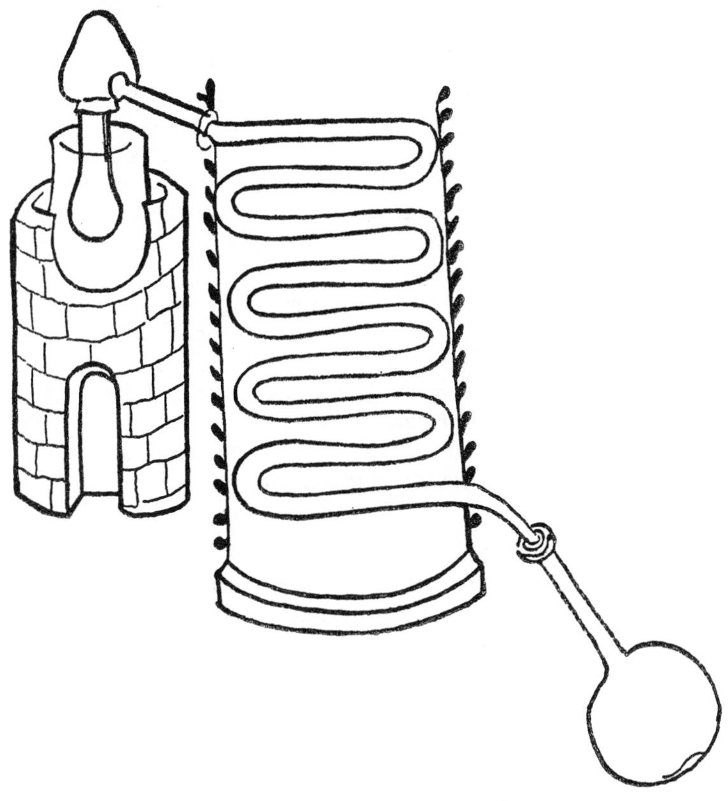

Figure 6. A Water-cooled Still

with a coiled tube immersed in a trough through which a constant flow of cold water was maintained (figure 6). It has been estimated that in this way an alcohol of 90 per cent purity could be obtained.

Alderotti's invention led to the discovery of rapidly increasing numbers of new substances, and particularly encouraged further applications of the mineral acids, though as yet only on a laboratory scale. It also led to an even wider diversity of forms of still, the principle of which is explained by a later alchemist, Giambattista della Porta (1538?–1615), as follows:

Both the vessel and the receiver must be considered according to the nature of the things to be distilled. For if they be of a flatulent nature and vaporous they will require large and low vessels and a more capacious receiver; or when the heat shall have been raised up to the flatulent matter and that find itself straitened in the narrow cavities, it will seek some other vent and so tear the vessel to pieces (which will flie about with a great bounce and crack, not without endamaging the standersby) and being at liberty will save itself from further harm. But if the things be hot and thin you must have vessels with a long and small neck. Things with a middle temper require vessels of a middle size. All which the industrious artificer may easily learn by the imitation of nature, who hath given angry and furious creatures as the lion and bear thick bodies and short necks; to show that flatulent humours would pass out of the vessels of a large bulk and the thicker part settle to the bottom; but then the stag, the ostrich, the camelopard [giraffe], gentle creatures and of thin spirits, have slender bodies and long necks; to show that thin and subtile spirits must be drawn through a much longer and narrower passage and be elevated to purify them.

There were few deliberate attempts at fractional distillation until the seventeenth century, but earlier alchemical literature has occasional passages indicating that the process was not unknown. Geber, for instance, whose Latin works appeared about 1300, has the following sensible observations to make (in the English translation by Richard Russell, 1678):

We will shew you the methods of distillation, with their causes. Therefore of that which is made by ascent, there is a twofold way or method. For one is performed in an earthen pan full of ashes; but the other with water in its vessel, with hay or wool, orderly so disposed, that the cucurbit, or distillatory alembeck, may not be broken before the work be brought to perfection [figure 7]. That which is made by ashes, is performed with a greater, stronger, and more acute fire; but what is made by water, with a mild and equal fire.

Figure 7. A Water-bath. (From The *Works of Geber*, Englished by R. Russell, ed. E. J. Holmyard, by courtesy of J. M. Dent & Sons Ltd)

For water admits not the acuity of ignition, as ashes doth. Therefore, by that distillation, which is made in ashes, colours and the more gross parts of the earth are wont to be elevated; but by that which is made in water, the parts more subtile, and without colour, and more approaching to the nature of simple wateriness, are usually elevated. Therefore more subtile separation is made by distillation in water, than by distilling in ashes. This he knows to be true, who, when he had distilled oyl by ashes, received his oyl scarcely altered into the recipient; but wishing to separate the parts thereof, was by necessity forced to distil it by water. And then by reiterating that labour, he separated the oyl into its elemental parts; so that from a most red oyl, he extracted a most white and most serene water, the whole redness thereof remaining in the bottom of the vessel.

Sublimation as a method of purifying such substances as sulphur and white arsenic was widely employed, and emphasis was laid upon preliminary tests to discover whether it should be carried out with strong, moderate, or gentle heat. Figure 8 shows one type of sublimatory. The furnace has an iron bar running transversely through it, about five or six inches from the bottom, and the glass vessel containing the substance to be sublimed is supported so that it nearly but not quite touches the bar. The perforated disk is arranged over the neck of the flask and helps to keep it in position, the holes in the disk acting as exits for hot gases from the furnace. The conical receiver fits over the mouth of the flask and serves to collect the sublimate. When it was necessary to ascertain if the sublimation was complete, an earthenware rod about the diameter of the little finger, and hollow for the lower half of its length, was held above the substances under experiment; if any solid formed in the hole the sublimation was regarded as unfinished. It is a matter of interest that during the alchemical period both benzoic acid and succinic acid were discovered by sublimation, from gum benzoin and amber respectively; benzoic acid was first described by Michel de Nostredame (Nostradamus) in 1556, and succinic acid by Agricola in 1546. In an apparatus used by Jean Béguin early in the seventeenth century for the sublimation of benzoic acid from gum benzoin, the receiver consisted of a sheet of grey paper folded double and twisted into the shape of a cone.

From contemporary drawings and paintings, we receive a strong impression that alchemists' laboratories were seldom characterized by tidiness, but there were exceptions. Thus the German alchemist Andreas Libavius (1540–1616) clearly had a propensity for neatness; he was an accomplished experimenter and several discoveries stand to his credit. He was the first to prepare stannic chloride and ammonium sulphate – the latter now produced in millions of tons annually for agricultural purposes – and observed the blue colour imparted to solutions of

Figure 8. A Sublimatory. (From The *Works of Geber*, Englished by R. Russell, ed. E. J. Holmyard, by courtesy of J. M. Dent & Sons Ltd)

copper compounds by ammonia; he also invented various wet and dry methods of analysis. In his book *Alchemia*, published in 1597, he gives a plan and elevation of his ideal 'chemical house'. This contained a main laboratory, doubtless housing the array of furnaces that still played so prominent a part in alchemical operations, a store-room, a preparation-room, a room for sand-baths and water-baths, a crystallizing-room, a room for the laboratory assistants, a fuel-store, and a wine-cellar. The last amenity was probably intended not so much for the promotion of conviviality among the staff as for storing the cheap wines from which alcohol was distilled.

In the principal laboratory, apparatus was arranged round the walls; besides the furnaces there were descensories, sublimatories, stills, crucibles, mortars and pestles, phials, flasks, and basins. It would doubtless have won the warm approval of the great Persian alchemist Razi, the contents of whose laboratory are described later in this book (p. 89).

References to temperature in alchemical literature are inevitably vague – thermometers were not invented until the early eighteenth century. The alchemists realized that a water-bath could not be made as hot as an ash-bath, but apart from such obvious distinctions the only 'fixed points' were such indefinite criteria as red-hot, white-hot, or as hot as a manure-heap (often used for gentle heating) or a broody hen. As late as 1622 the German alchemist J. D. Mylius recognized only four degrees of heat, namely that of the human body (which is fairly constant), that of June sunshine, that of a calcining fire, and that of fusion.

Some idea of the cost of equipping a laboratory in England at the beginning of the seventeenth century can be gained from a manuscript preserved at Alnwick Castle and quoted by J. W. Shirley. It refers to the fitting-out of a still-house (distilling-room) in the Tower of London for the 'Wizard Earl', Henry Percy, ninth Earl of Northumberland, who was imprisoned there in 1606–7 with Sir Walter Ralegh, Lord Cobham, and Lord Grey. Both the Wizard Earl and Ralegh were interested in alchemy. The accounts show that bricks, tiles, and other necessaries for making the furnace cost 12s 6d, while the carpenter was paid 22s 10d for timber for the bench, shelves, and stairs. Two lead cisterns cost 15s 10d, two stills 8s 6d, a copper vessel 7s 7d, bellows 10d, a file 3d, a steel chisel 4d, a pair of long compasses 4d, various glass-ware 31s 6d, and 'diverse other necessaries as Colebasketts handbucketts syves payles potts pannes packe threed tape wyer Strayners treys pastbord wax paper matts Candlesticks postage and such like' 22s 6d. Further expenses a year later included 11s 8d for a new tin still, 5s 11d for a pair of goldsmith's scales and a box to keep them in, 4s 4d for weights to go with the scales, 3s 8d for phials, 2s 1d for repairs to the walls and glazed windows, and 3s 10d for an iron pestle and mortar. These figures may be compared with the expenditure on alchemical equipment made by James IV of Scotland a hundred years earlier (p. 219).

5

ISLAMIC ALCHEMY

EVENTS of the seventh to tenth centuries proved to have a decisive effect upon the development of alchemy. Those years saw the foundation of the religion of Islam and the establishment of the empire of the Caliphs. Muhammad, the Prophet of Islam, was born about A.D. 570, and, being left an orphan at an early age, gained his living by acting as a camel-driver for trade caravans. He belonged to the Hashim family of the tribe of Kuraish and lived at Mecca, which owed its prosperity to its situation on the caravan route from Abyssinia through the Yemen to Palestine and Syria. Though never of robust health, Muhammad had a reflective and sensitive mind, much given to meditations of a religious character, and he gradually became convinced that he had a divine mission to fulfil. His marriage to Khadija, a rich widow whose camels he had driven, enabled him to pursue his religious ponderings untrammelled with monetary cares, and gave him leisure for lengthy discussions with the numerous Jews and Christians who lived at Mecca or visited it for business.

Muhammad soon became deeply dissatisfied with the religion of his fellow-countrymen, or rather with their lack of it. At that time, indeed, the Arabs were little better than fetishists, worshipping stones, stars, trees, wells, and other inanimate objects, having no conception of prayer, and believing in evil or at least unfriendly spirits – the jinns or genies of the 'Arabian Nights'. The chief centre of worship was the Kaaba or 'cubical' building, in which was fixed a sacred black stone alleged to have fallen from heaven in or about the time of Abraham. This relic – still there – attracted pilgrims from far and wide, especially at the time of the great fair held at Mecca every year, and enterprising citizens were not slow to turn the circumstance to their financial advantage. Idols were added to the Kaaba according to the predilections of various tribes, who were thus led to the pilgrimage;

a ceremonial tradition grew up; and the leading families of Mecca, among whom many of the Kuraish were prominent, became wealthy, arrogant, and materialistic.

It was not until he was forty years old that Muhammad began to preach against such worldliness and polytheism, urging penitence and the worship of the One True God – in Arabic, Allah. His system was based on Judaism and Christianity, but he regarded Christ as a prophet like Moses or Isaiah and did not admit his divinity; he looked upon himself as the last and greatest of the prophets; and his religion, Islam, signifies 'submission to the will of God'. Understandably enough, the Kuraish and other Meccan plutocrats heartily disapproved of the new teaching, which was calculated to wreck the customs upon which their wealth largely depended, and they made conditions so unpleasant for Muhammad that he decided to migrate from Mecca. With some of his followers he moved northward to the city of Yathrib, which afterwards became known as Medina, or, in full, Medinat al-Nabi, 'the city of the Prophet'. Here his fortunes flourished and he soon became the leader of a sizable and enthusiastic religious community. The Muhammadan era dates from the year of this migration (*hijra* or hegira), namely 622, but uses lunar years of twelve months of alternately 30 and 29 days with occasional intercalation of one day at the end of the twelfth month; the average length of the Muslim year is thus about 354 days 9 hours.

The Prophet's activities at Medina marked the beginning of Islam not only as a religion but as a religious state, and by 630 he felt strong enough to re-enter Mecca and subdue it. When the city surrendered to him he behaved generously to its defenders, but immediately set about destroying the idols and all traces of polytheism: 'There is no god but God, and Muhammad is the Apostle of God' became and remains the watchword of his adherents, the Muslims. To this strict monotheism Muhammad added the principles of belief in the mercy of God, the necessity for the repentance of sinners, the obligation of regular prayer and fasting, the duty of alms-giving, and the incumbency of making a pilgrimage to Mecca at least once in a lifetime.

Though the Prophet never owned sway over the whole of

Arabia he welded together many of its quarrelsome tribes and factions and thus provided what was to be a powerful weapon in the conquests soon to follow. The expansion of Islam has sometimes been imagined as a *jihad* or holy war, with the Muslims surging forward carrying a sword in one hand and the Koran in the other, but this is a travesty of the truth. The time was in fact ripe for an Arab war of conquest; the nation was undergoing a period of restless and energetic life and had been seized with the urge to leave its barren and inhospitable deserts for the richer lands so temptingly near. The material resources of Arabia were small and the population was increasing; hunger and poverty were becoming unbearable, and war was the only way out.

Sporadic raids across the frontiers of the Byzantine (Eastern Roman) Empire took place even before the Prophet's death in 632, and in 635 the Arabs captured Damascus. Yet in these early affrays comparatively few Muslims took part; the brunt of the fighting was borne by the hardy and heathen Bedouin of the desert, more anxious for spoil than to proselytize for Islam – of which the majority of them had probably scarcely heard. But the Faith was rapidly gaining new adherents, and when the Arabs beseiged Jerusalem in 636 they were led by the Muslim Omar.

Muhammad had named no one to take his place as leader of Islam, but when he died Abu Bakr, father of his favourite wife Ayesha, was chosen as Caliph ('successor') by popular acclamation. Abu Bakr was succeeded in 644 by Omar I, the victor of Jerusalem, who was the first ruler to assume the title of *Amir al-Mu'minin* or Commander of the Faithful. Thenceforward Islam played a leading part in Arab affairs, and under Omar and the early caliphs who followed him successful wars of conquest brought Egypt, Palestine, Syria, much of Asia Minor, Crete, Sicily, Rhodes, Cyprus, and part of North Africa under Muslim control. Progress in Africa was halted for a time by the stiff resistance of the Berbers, but early in the eighth century Muslim armies occupied Ceuta (ancient *Septem*) and crossed into Spain at the Pillars of Hercules. Their commander here was Tarik, after whom Gibraltar (Mountain of Tarik, *Jabal Tarik*) was named. The invasion of Spain was quickly completed and the conquerors then passed into France; here, however, they were finally brought to a halt in 732, when Charles the Hammer dealt them a crushing blow at Poitiers.

Thus within a century after the Prophet's death Islam had become a vast empire stretching from the Pyrenees to the Indus, and many and various races had become incorporated into its civilization. Some of these peoples continued to speak their own languages, but Arabic was the religious, official, and literary language throughout the empire, and was occasionally made compulsory for speech in public. A ninth-century Bishop of Cordoba complained that his fellow Christians in Spain read Arabic poetry and romances not in order to be able to refute Muslim theologians but so that they might express themselves more elegantly and correctly in the Arabic language: 'there is hardly one among a thousand to be found who can write to a friend a decent letter in Latin'. This general adoption of Arabic throughout the educated classes in Islam accounts for the fact that numerous Muslim works on alchemy are Arabic only linguistically, their authors being of Persian or other nationality and not Arabs. The most distinguished alchemist of Islam, Jabir (p. 68), was however a member of a well-known Arab tribe.

When political conditions became quieter the Muslims soon manifested a great interest in learning, and having overrun not only Alexandria and Harran but all the other principal centres of Greek culture they were able to indulge it to the full. Under such enlightened rulers as Harun al-Rashid (764?–809) and Al-Ma'mun (786–833) large numbers of academies and observatories were set up and the chief works in Greek on philosophy, astronomy, mathematics, medicine, and other sciences were translated into Arabic, for the most part by Syriac-speaking Nestorians. From the eighth century onwards Islam was producing scholars of her own.

KHALID IBN YAZID

According to Ibn al-Nadim, a biographer of the second half of the tenth century, the first Muslim to interest himself in alchemy was the Umayyad prince Khalid ibn Yazid, who died about 704. Ibn al-Nadim says that Khalid had a general love for the sciences but was particularly attracted to that of alchemy; so he ordered some Greek philosophers to be summoned from Egypt and instructed them to translate alchemical books from the Greek and Coptic languages into Arabic. 'This,' adds the biographer, 'was the first translation from one language to another in Islam.'

Other historians elaborate this brief sketch. Khalid was the son of the Caliph Yazid I, who died in 682 and was succeeded by Khalid's elder brother Muawiya II. Muawiya, however, survived his father by only a few months, and the Caliphate should then have passed to Khalid. But at that time Khalid was still in his teens and was considered too young to rule, so his relative Marwan was made Caliph on the understanding that Khalid was to be next in succession. Once in power, Marwan went back on his pledge and appointed his own son Abdul-Malik as successor, at the same time accusing Khalid's mother of immorality. The lady retaliated by causing Abdul-Malik either to be poisoned or to be suffocated with pillows while he slept, and these events so sickened the young prince that he withdrew from court life and devoted his remaining years to the study of the sciences.

Khalid is said to have studied alchemy under a Christian scholar of Alexandria, one Marianos or Morienus, who had himself been a disciple of the earlier Alexandrian alchemist Stephanos (p. 29). The story goes that Khalid had previously surrounded himself with self-styled experts in the art but was invariably disappointed at their failure to effect transmutation. Morienus, who was leading the life of a hermit at Jerusalem, heard of Khalid's great interest in learning and resolved to pay him a visit in the hope of converting him from Islam to Christianity. He was received with courtesy, and finding that Khalid desired above all things to witness the alchemical production of gold he asked for a room and equipment and there and then performed a successful transmutation. When Khalid had gazed at the alchemical gold, with true Oriental despotism he ordered the execution of the fraudulent alchemists, and in the resulting commotion Morienus vanished.

Several years elapsed before Ghalib, one of Khalid's most trusted servants, learned from another hermit where Morienus was to be found, and was despatched immediately to bring him back. On this second occasion of their meeting, Morienus, though still expressing the hope that God would convert Khalid to a better mind, revealed the secrets of alchemy and answered the many questions that were put to him. What happened to Morienus afterwards we are not told, but Khalid set to work to enshrine his newly acquired knowledge in a number of alchemical poems. Some verses ascribed to Khalid are quoted by later writers, and a collection of similarly ascribed alchemical poems is preserved in a library at Istanbul. Ibn al-Nadim says that he had himself seen works of Khalid bearing the following titles: 'The Book of Amulets'; 'The Great and Small Books of the Scroll'; and 'The Book of the Testament (to his son) on the Art'. The most celebrated of the works supposedly written by Khalid is entitled 'The Paradise of Wisdom' which according to a Muslim biographer, Hajji Khalfa (1599–1658), contained 2315 verses.

Before inquiring into the amount of truth contained in the above stories, we may complete the picture of an engaging character by adding that, as a youth, Khalid forcibly ejected a rioter from the pulpit of the Church of St John the Baptist at Damascus (afterwards the Great Mosque); and that he appears to have been instrumental in the decision that state accounts should be kept in Arabic instead of, as had been the custom, in Persian.

The question of the historicity of Khalid's preoccupation with alchemy was considered in great detail by the German scholar Julius Ruska, whose work we shall often have occasion to mention. Ruska's intensive study of Muslim alchemy so often revealed examples of falsified history, of obvious legends taken seriously, and of later writings fathered upon earlier authors who could not possibly have written them, that in the end he developed what we may think an exaggerated and unreasonable scepticism concerning the authorship of any early Arabic alchemical work. It is true that the subject bristles with difficulties, and undeniable that many spurious books and stories gained currency. At the same time, Muslim tradition and history can often be confirmed by both internal and external evidence, and Stapleton and others have shown that Ruska's disbelief has in some important respects been unfounded. It may be a more scholarly attitude to take nothing as true that cannot be proved, but when one is dealing with such an obscure subject as alchemy, in an alien civilization of centuries ago, such a method of attack is not very fruitful. With due reservation and caution, more progress can be made by paying respect to accepted beliefs until they can clearly be shown untenable. It is unnecessary to remark that a procedure of this kind is inapplicable to such matters as circumstantial accounts of alchemical transmutations, where there is an overwhelmingly strong *a priori* reason for incredulity.

In the case of Khalid, that he was an Umayyad prince living at Damascus about 660–704, and that he failed to succeed to the Caliphate, are unquestioned historical facts. Ruska, however, casts doubt upon the likelihood that a young man of princely rank should interest himself in alchemy. Yet the story is well authenticated, and other examples of royal attention to alchemical matters are not far to seek – we may remember the Emperor Herakleios and James IV of Scotland. Neither does it seem improbable that, if Khalid wanted information on the subject, he should get it from a monk of Alexandria, then and for hundreds of years earlier one of the chief centres of alchemical lore. Ruska is on surer ground when he pronounces the extant works ascribed to Khalid to be, at least for the most part, pseudepigraphical, for while a historian of 950 knew of only three alchemical verses said to have been composed by Khalid, Hajji Khalfa, seven centuries later, knew of 2315 (p. 65). Even in these circumstances, however, caution would seem advisable, for many works passing under Khalid's name still lie unexamined in the great libraries of India, Egypt, and Europe – and it is only a few years ago that a hitherto unknown work by Chaucer was discovered in a Cambridge library, nearly six centuries after it was written.

ORIGINS OF ALCHEMY IN ISLAM

We shall hear of Khalid again (p. 106), this time in connexion with the introduction of alchemy to western Europe; meanwhile we may leave him to inquire more closely into the sources of Muslim alchemy. Whether he was in fact the pioneer of alchemical knowledge in Islam must remain dubious, but one feature of his story at least is authentic – namely that alchemy came to the Muslims originally from Alexandria. The names venerated by the early Arabic alchemists are those already familiar to us, such as Hermes, Agathodemon, Plato, Zosimus, Democritus, Heraclius, Ostanes, Stephanos, Apollonius, Alexander, Archelaos, Mary the Jewess, and many others, the main interest of the list lying in the evidence it provides that Islam appropriated the Greek alchemical authorities *in toto*. Confirmation, if any were needed, of the close affiliation between Greek and Arabic alchemy is provided by the large number of Greek technical terms transliterated into Arabic from Hellenic treatises.

The transmission was made chiefly through direct contact in Alexandria and other Egyptian cities, but partly by intercourse with the intellectual centres of Harran, Nisibin, and Edessa in western Mesopotamia. This subsidiary channel helps to explain the unmistakable traces of Persian and even Assyrian influence in Muslim alchemy, manifested by linguistic affinities in technical terms and usages and in names of minerals; thus *abaru*, a name that occurs very frequently in Arabic treatises and signifies the metal (lead or antimony) extracted from collyrium, is the Assyrian word for the same substance. Still another point of contact between the early Muslims and the body of established alchemical doctrine was the celebrated academy at Jundi-Shapur in south-west Persia, which was still flourishing at the time of the Abbasid caliphs Harun al-Rashid and Al-Ma'mun.

It has already been mentioned (p. 63) that Nestorian Christians played a great part in translating Greek works into Arabic, and doubtless the first Muslim alchemists no less than their fellow-students of other branches of learning were indebted to Nestorians in this way. The greatest of these translators was Hunain ibn Ishaq, who was born in Hira (southern Iraq) in

809–10 and died at Baghdad in 877. Though his main concern was with the establishment and translation of Greek medical texts, he translated other works as well, and his fame became such that the Caliph Mutawakkil created or endowed a college where scholars were employed to make translations under Hunain's supervision. In view of the close connexion that existed between medicine and alchemy it is probable that Arabic versions of some Greek alchemical works passed through Hunain's hands if they were not actually made by him.

Syrian pagans from Harran were also widely employed in translation; they were star-worshippers and diligent astrologers. These Sabians,* as the Arabs called them, possessed exceptional skill as linguists, and the ease with which they acquired Arabic recommended them to the court at Baghdad, where they were tolerated in spite of their unbelief. But this dependence on foreign translators was not long complete; as early as the eighth century there were Muslim scholars who could read Greek, and the number increased with the lapse of time.

JABIR IBN HAYYAN

Like printing, which reached its highest pitch of perfection while still in its infancy, Islamic alchemy never surpassed the level it attained with one of its earliest exponents, Jabir ibn Hayyan. Jabir has long been familiar to Western readers under the name of Geber, which is the medieval rendering of the Arabic word, but it is only recently that serious research has been carried out on his life and work. At the present time, however, owing to the investigations of H. E. Stapleton, J. Ruska, P. Kraus, the present writer, and others, it is possible to draw some picture of his circumstances and to form a fairly close idea of the part he played in the development of alchemy. Much is still conjectural, but the following account is probably authentic in the main.

Jabir ibn Hayyan means Jabir the son of Hayyan; and Jabir was often further designated Al-Azdi, implying that he belonged

*Not to be confused with the Sabaeans, inhabitants of the pre-Islamic Kingdom of Sheba (in the present Yemen).

to the Azd tribe of south Arabia, and Al-Kufi, to indicate that he was a native or citizen of Kufa. Almost as often he is described as Al-Tusi, or the man of Tus in Khorassan, and Al-Sufi, which signifies that he was a member of the community that cultivated a species of mysticism known as Sufiism. These various designations, confusing as they seem at first sight, were in fact clues that did much to clear up uncertainties about Jabir's life. We may deal with them in turn, starting with Al-Azdi and Al-Kufi.

In 638 the Caliph Omar was visited at Medina by a deputation of Arabs from Al-Medain, a town on the Tigris that they had recently taken by assault. The Caliph was startled by their sallow and unhealthy look, and asked the cause of it. They replied that the climate of the town was unwholesome and that they could not accommodate themselves to it. The Caliph therefore ordered inquiry for some more salubrious and congenial spot. A plain on the western bank of the Euphrates was finally chosen, and there the city of Kufa was founded. The new town suited the Arabs well, and to it they migrated in large numbers. But the dwellings were at first made of reeds, and fires were frequent; so after a particularly disastrous conflagration the city was rebuilt of less inflammable material, and the streets were laid out in regular lines. In orderly fashion, befitting what was originally a garrison town, the various Arab tribes were settled in particular quarters of the city, the arrangement no doubt serving also to lessen the possibility of civil commotion. At its heyday, Kufa numbered some 200,000 citizens and was one of the principal cities of the Umayyad Caliphate.

One of the tribes whose members were present at Kufa in sufficient numbers to be assigned a definite quarter was that known as Al-Azd, and Jabir's description as Al-Azdi al-Kufi thus seemed to indicate that he belonged to that part of the Azd tribe domiciled at Kufa. Now a certain Azdi called Hayyan, a druggist of Kufa, is mentioned in Muslim chronicles in connexion with the political machinations that, in the eighth century, finally resulted in the overthrow of the Umayyad dynasty by that of the Abbasids. By 719–720 popular feeling against Umayyad misrule had become widespread and in those years was begun an underground movement in favour of the Abbasid family led by the Imam Muhammad ibn Ali. The movement was supported by the Shi'ite sect, which held that succession to the Caliphate was the right only of one descended from the Prophet by way of Ali, his cousin and son-in-law, and Fatima, Ali's wife and the Prophet's daughter; but there now being no one so qualified the Shi'ites decided to lend their support to the movement to establish the Abbasids, who were descended from the Prophet's uncle Abbas.

The first of the Shi'ites who came forward included Abu Ikrima, a saddler, and Hayyan, the druggist, who went to see Muhammad ibn Ali at a rendezvous in Syria. The Imam sent them to Khorassan with instructions to invite people to swear allegiance to the Abbasid cause and to stir up discontent against the Umayyads on account of their evil conduct and grievous tyranny. Many in Khorassan responded to their call, but their activities became known to the governor of the province, so he sent for them and asked them to account for themselves. 'Who are ye?' he said. 'Merchants,' they replied. 'And what,' said he, 'is this which is currently reported concerning you?' 'What may that be?' they asked. 'We are informed,' said he, 'that ye be come as propagandists for the house of Abbas.' 'O Amir,' they answered, 'we have sufficient concern for ourselves and our own business to keep us from such doings.' So he let them go, and they were able to continue their subversive activities for another two years; then, however, they were once more apprehended and this time beheaded and their bodies impaled. They and their fellow-conspirators had nevertheless done their work, for within a few years the house of Umayya had been overthrown and the Abbasids reigned in their stead. The change did not in fact turn out as the Shi'ites had hoped, and once the new dynasty was established many of them were slaughtered.

In the course of his clandestine wanderings in Khorassan Hayyan must have visited most if not all the important towns there, one of which was Tus, near the modern Meshed. If we assume that Jabir was born at Tus, as he might have been about 721 or 722, he could rightly be called Al-Tusi.

Of Jabir's common appellations we are thus left with Al-Sufi.

Sufiism is an ascetic system of mysticism within Islam, and was so called from the Arabic *suf*, wool, because the early adherents used to wear a coarse garment made of this material. Partly a reaction from the extravagance and licence of the court, it soon became a movement whose members sought ecstatic union with God, practised rigid austerity, and gave themselves to contemplation and religious exercises. Many of its tenets were similar to those of Neo-Platonism, by which indeed it was much influenced – a fact not without significance in the history of alchemy.

The fatherless boy Jabir was sent to Arabia, perhaps to some of his kinsmen of the Azd tribe, to be cared for until he could fend for himself; presumably he would have been with Bedouin so as to speak the pure language of the Koran, urban accents and dialects being frowned upon by the educated. In one of his books Jabir tells us that while in Arabia he studied the Koran, mathematics, and other subjects under a scholar named Harbi al-Himyari, but we have no further information about his early life; if his father's business had been kept on at Kufa he might have returned there for a time and so have begun his acquaintance with chemical operations. Later on, he certainly lived at Kufa for many years.

Jabir first emerges as a definite figure in middle life, when we find him established as alchemist at the court of Harun al-Rashid and as the personal friend of the sixth Shi'ite Imam Ja'far al-Sadiq (700–65). His friendship with the Imam – a much revered figure – is understandable if Ja'far remembered Hayyan's sacrifice, unfortunate as the event for which he died proved eventually to be for the Shi'ites. Jabir often refers with respect and affection to his 'Master', Ja'far, who was a man of culture and learning and whose conversation must have been a great intellectual stimulus to the younger one.

Jabir was also in favour with the Barmecides, the Caliph's all-powerful ministers, some of whom figure in 'The Thousand Nights and a Night'. Through the medium of the vizier Ja'far the Barmecide, Jabir was brought into contact with the Caliph himself, 'for whom he wrote a book on the noble art of alchemy entitled "The Book of Venus". In it he described wonderful experiments of a very elegant technique.' On one occasion we find him accompanying his patrons to the slave-market to buy handmaids, while on another he describes a cure he affected in Yahya the Barmecide's household. One of Yahya's concubines, unequalled in beauty and perfection and deportment and intelligence and accomplishments, was at the point of death from some obscure disease. Jabir was called in, and, says he,

I had a certain elixir with me, so I gave her a draught of two grains of it in three ounces of vinegar and honey, and in less than half an hour she was as well as ever. And Yahya fell at my feet and kissed them, but I said, 'Do not so, O my brother.' And he asked about the uses of the elixir, and I gave him the remainder of it and explained how it was employed, whereupon he applied himself to the study of science and persevered until he knew many things; but he was not so clever as his son Ja'far.

It is said that it was through the efforts of Jabir that the second importation of Greek scientific works from Byzantium was made, the first being that made under the auspices of Khalid ibn Yazid some three-quarters of a century earlier. We need not take this too literally, for, once begun, importation went on without interruption; but it is quite likely that Jabir's thirst for learning urged him to get the process expedited. Though his main interests lay in alchemy he was a widely read scholar and may have had some knowledge of Greek. His own list of his writings, which has come down to us at second hand from Ibn al-Nadim, shows that he wrote books on a wide variety of subjects – a fact that need occasion no surprise in view of the vast extent of the intellectual treasures now becoming available to the Muslim world. Thus besides very numerous books on alchemy he composed a book of astronomical tables, a commentary on Euclid and another on the 'Almagest' of Ptolemy, several books on talismans according to the opinions of Apollonius of Tyana, and many others on such widely differing topics as philosophy, logic, medicine, automata, military engines, magic squares, and mirrors. It should be remembered that many of these 'books' were very short, being little more than what we should now describe as 'articles' or 'papers'; but their very multiplicity set in train doubts as to their authenticity – a topic to which we shall return shortly. A great number of the works mentioned still exist in manuscript, but only a comparatively small fraction of them has yet been studied.

When Jabir first went to Baghdad we do not know; neither is it certain how long he stayed there. He had a laboratory at Kufa, which was rediscovered, about two centuries after his death, during the demolition of some houses in the quarter of the town known as the Damascus Gate. In it was found a golden mortar weighing two and a half pounds, of which the royal chamberlain took ceremonious possession. The laboratory was otherwise empty, though that it had been designed for the performance of various chemical operations was clear from the layout.

In 803 Harun al-Rashid finally tired of the Barmecides, who had grown so powerful as to be a continual menace to him, and executed one of them and banished the rest. Jabir, we are told, was involved in the disgrace of his patrons and found it prudent to return to Kufa, where he spent the remainder of his life in seclusion. According to one authority he survived until the days of Al-Ma'mun, who ruled from 813 to 833; another says that he died at Tus in 815, with the manuscript of one of his works, 'The Book of Mercy', under his pillow.

Turning now to a consideration of the writings that pass under Jabir's name: there are so many of them that it seemed hardly possible for one man to have written them all, and the suspicion grew that some at least were fathered on him by later writers. The investigations of Kraus, published in 1942–3, showed that the suspicion was well founded. Critical examination of the Jabirian *corpus* proved beyond doubt that much of it must have been expanded or perhaps originally written by members of the Isma'ilite sect, which arose in the tenth century and rapidly won large numbers of adherents; it is still in existence. One of its main tenets was that Muhammad was not the last of the prophets but only one of a series; another was that all religions had some elements of truth, the Isma'ilite system embracing the whole of them. A notorious branch of the sect comprised the Assassins, so called from its members' habit of drugging themselves with hashish before setting out to wreak vengeance on those who had incurred their leader's displeasure.*

From internal evidence, which often reveals indubitable evidence of Isma'ilite influence, it is clear that in very many cases books ascribed to Jabir cannot have been written by him in the form in which they have come down to us. How much of their contents is the genuine work of Jabir of the eighth century, and how much is material added in the tenth, there is no obvious means of deciding; we can say that such and such a passage must be tenth century, but not that such and such another passage must be eighth century. In the account that follows we shall use the word 'Jabir' to imply the Jabirian *corpus* as a whole.

The most important groups of treatises in the *corpus* are (in probable chronological order): (i) 'The Hundred and Twelve Books'; (ii) 'The Seventy Books'; (iii) 'The Ten Books of Rectifications'; and (iv) 'The Books of the Balances'. Some of the first group are dedicated to the Barmecides, and the group as a whole is ultimately based on the 'Emerald Table' of Hermes (p. 97). The second group is interesting inasmuch as it, or a good deal of it, was translated into Latin in the twelfth century by Gerard of Cremona. The third group purports to describe the alchemical advances made by various alleged alchemists including Pythagoras, Socrates, Plato, and Aristotle; and the fourth group develops Jabir's theory of the 'balance', to which reference is made below.

On the constitution of matter, Jabir held the Aristotelian conception of the four elements: fire, air, water, earth: but developed it on different lines. He postulated first the existence of four elementary qualities or 'natures', namely hotness, coldness, dryness, and moistness. When these natures united with substance they formed compounds of the first degree, namely hot, cold, dry, moist. Union of two of these gave rise to fire (hot + dry + substance); air (hot + moist + substance); water (cold + moist + substance) and earth (cold + dry + substance). In metals, two of the 'natures' are external and two internal, a point to

*The present head of the Isma'ilite sect is the Aga Khan, who is a descendant of the leader of the Assassins known to the Crusaders as 'The Old Man of the Mountains'.

which further reference is made later (p. 77). Thus in his 'Seventy Books' Jabir says that lead is cold and dry externally and hot and moist internally; gold, on the other hand, it hot and moist externally and cold and dry internally.

He believed that, under the influence of the planets, metals were formed in the earth by the union of sulphur (which would provide the hot and dry 'natures') and mercury (providing the cold and moist). This theory, which appears to have been unknown to the ancients, represents one of Jabir's principal contributions to alchemical thought; it may have been wholly original, though perhaps Jabir found the germs of it in Apollonius of Tyana. It was generally accepted by later generations of alchemists and chemists, and survived until the rise of the phlogiston theory of combustion in the concluding years of the seventeenth century.

A proviso should be made concerning the character of the sulphur and the mercury of which Jabir supposed metals to be formed. He knew quite well that when ordinary sulphur and mercury are heated together the product obtained is a non-metallic stony substance; in fact he describes this very experiment and says that the resulting solid is cinnabar. The sulphur and mercury composing metals were, then, not the substances commonly known by those names, but hypothetical substances to which ordinary sulphur and mercury formed the closest available approximations (cf. p. 94).

The reasons for the existence of different kinds of metal are that the sulphur and mercury are not always pure, and that they do not always unite in the same proportion. If they are perfectly pure, and if also they combine in the most complete natural equilibrium, then the product is the most perfect of metals, namely gold. Defects in purity and, particularly, proportion result in the formation of silver, lead, tin, iron, or copper; but since these inferior metals are essentially composed of the same constituents as gold, the accidents of combination may be rectified by suitable treatment. Such treatment, according to Jabir, is to be carried out by means of elixirs.

He was convinced that to try to effect transmutations empirically was a waste of time; he believed that order reigned in the material world, and that qualitative changes in substances could be explained on a quantitative basis. He was thus led to his characteristic conception of the 'balance', which he elaborated particularly in his 'Books of the Balances' and which seems to have been one of his later theories. By 'balance' he did not refer to equality of mass; in fact, although he does describe a hydrostatic balance in one of his books, the importance of comparing masses was as little perceived by him as by all other alchemists. His balance was an equilibrium of 'natures', and a great deal of his work was devoted to attempts to establish the equilibrium figures in gold – the perfect metal – so that the same balance could be effected in base metals, thus bringing about their transmutation.

It would not be profitable to follow the development of such a scheme in detail, but it is of interest to examine one of Jabir's lines of attack. This was done by Kraus and Stapleton, on whose work the following account is principally based. We find throughout Jabir's works a series of numbers to which he attaches great importance: it is 1, 3, 5, 8 (totalling 17), and 28. He says that everything in the world is governed by the number 17 – metals, for instance, have 17 'powers'. Now the numbers composing the total of 17, namely 1, 3, 5, 8, are a significant part of those of the magic square of the first nine digits:

4	9	2
3	5	7
8	1	6

Figure 9, Magic Square of the First Nine Digits Gnomonically Analysed

Here the total is 45, but analysing the square gnomonically, as indicated by the heavy lines in the figure, gives 1, 3, 5, 8, in the remaining square, while the total in the gnomon is 28. This magic square, which was known to the Neo-Platonists of the third century and is doubtless much older (p. 20), is thus clearly the source of Jabir's significant numbers. It will be remembered that the list of his books includes one on magic squares, and this particular square had associations for the Sufi mystical society of which he was a member.

One way in which Jabir used these numbers was in the application of alphabetical numerology to elucidate the constitution of the metals. Each of the four elementary qualities or 'natures' was supposed by him to have four degrees and seven subdivisions, giving a total of 28 × 4, i.e. 112, 'positions'. The letters of the Arabic alphabet, 28 in all, were assigned to the subdivisions of heat, coldness, dryness, and humidity, and the scheme was extended to the values of the four degrees according to the series 1, 3, 5, 8. The degrees and subdivisions were equated to weights on the Arabic system of 2 qirats = 1 danaq, 6 danaqs = 1 dirham, and a table was constructed in which, for example, the letter *b* denoted, in the second degree of coldness, a weight of 3½ dirhams; in the fourth degree *b* corresponded to a weight of 9⅓ dirhams. The remaining letters were similarly calibrated.

To determine the 'balance' of lead on this scheme, its name was analysed arithmo-alphabetically, only the consonants being used. Lead in Arabic is *usurb*, but the first letter, not indicated in the transliteration, is the consonant *alif*. *U* being a vowel, the operative letters are therefore *alif*, *sin* (*s*), *ra* (*r*), and *ba* (*b*). *Alif*, being the first letter of the name, represents heat of the first degree, and is equivalent to a weight of 7 danaqs; *sin* the second letter, represents dryness of the second degree, and is equivalent to 1 dirham; *ra* represents humidity of the third degree, 1¼ dirhams; and *ba* represents coldness of the fourth degree, that is 9⅓ dirhams. Hence a lump of lead weighing 12¾ dirhams would contain the above weights of heat, dryness, humidity, and coldness, and this composition would hold for any specimen of lead. The fact that another Arabic name for lead, namely *rasas*, gives an entirely different result may cause misgivings as to the reliability of this method of analysis.

It was mentioned earlier that Jabir distinguished between the external and the internal composition of a metal. One reason for this distinction can be found in the figures just elicited. Metals are composed of heat, coldness, dryness, and humidity, but there is a limiting condition: opposing 'natures' are in the ratio of either 1:3 or 5:8 or vice versa. The figures for lead, however, do not agree with these ratios, and the difficulty is even greater with *fidda*, silver, when analysed in the same way – it proves to consist merely of heat and coldness in equal proportions. Jabir was therefore forced to use a further hypothesis, namely that the analysis reveals only the peripheral constitution; the balance must be restored by the constitution of the interior. Hence for silver the total composition, external and internal together, must be arrived at by calculation:

Heat:
$1\frac{1}{2}$ *danaqs* + a complement of $5\frac{1}{2}$ *danaqs* $= 1\frac{1}{6}$ *dirhams*
Coldness:
$1\frac{1}{2}$ *danaqs* + a complement of $3\frac{1}{4}$ *dirhams* $= 3\frac{1}{2}$ *dirhams* $(= 3 \times 1\frac{1}{6})$
Humidity:
0 *danaqs* + a complement of $5\frac{5}{6}$ *dirhams* $= 5\frac{5}{6}$ *dirhams* $(= 5 \times 1\frac{1}{6})$
Dryness:
0 *danaqs* + a complement of $9\frac{1}{3}$ *dirhams* $= 9\frac{1}{3}$ *dirhams* $(= 8 \times 1\frac{1}{6})$

19[illegible] *dirhams* $(= 17 \times 1\frac{1}{6})$

The unit of $1\frac{1}{6}$ seems to be chosen here so that fractions of a dirham not represented by a danaq or a qirat are avoided.

The transmutation of one metal into another is thus an adjustment of the ratio of the manifest and latent constitutions of the first to those of the second, an adjustment to be brought about by an elixir. According to Jabir there are various elixirs suitable for specific transmutations, but transmutations of every kind can be brought about by a grand or master elixir. This grand elixir was itself of two grades, differing only in power, a point illustrated by the story related by an alchemist called Dubais ibn Malik and published by Stapleton. Dubais said,

I was living at Antioch, where I had settled, and there I had a friend who was a jeweller by profession, to whose shop I often resorted. Now, as we were talking together one day, a man came in, and, having saluted, took his seat. After a while he removed from his arm an armlet which he handed to my friend. It was set with four jewels, and an amulet of red gold was fitted into it. On the amulet was inlaid a clear inscription in green emerald, which read as follows: Al-Hakim bi-amrillah puts his trust in God [Al-Hakim bi-amrillah was ruler of Egypt 996–1020]. I was astounded at the fineness of the jewels, the like of which I had never before seen, nor had I ever thought to see the like in the world, and it occurred to me that this amulet must have been stolen from the treasury of Al-Hakim, or it might have fallen from his arm, and this man had picked it up, since such jewels are to be found only in the treasuries of kings, or among their heirlooms.

It was finally purchased by Dubais for 3000 dinars. Inside the amulet was found a manuscript, pronounced by Dubais, who was acquainted with the shaky handwriting of Al-Hakim, to be in the holograph of that king, containing an account of two ways of making the Red Elixir, according to the method of Moses and the rest of the Prophets as handed down by the Imam Ja'far al-Sadiq. Dubais was successful in carrying out the operations, both of the Lesser Way, whereby an Elixir was made capable of converting 500 times its own weight of base metal into gold, and of the Greater Way, whereby an Elixir was prepared of which only one dirham was required for the conversion of 3000 dirhams of base metal.

The Alexandrian and Harranian alchemists preferred, if they did not exclusively use, mineral substances in their attempts to prepare elixirs for transmutation, but Jabir was an innovator and introduced both animal and vegetable products to the alchemical armoury. Among the former he mentions the marrow, blood, hair, bones, and urine of lions, vipers, foxes, oxen, gazelles, and donkeys both domesticated and wild. Suitable plants included aconite, olive, jasmine, love-in-a-mist, onion, ginger, pepper, mustard, pear, and anemone.

Such lists provide an indication that Jabir was more than a theorizer, and there is indeed much in his books to show that he was well versed in chemical operations. Though his theory is complex, obscure, and often to our modern minds ridiculous, he can be perfectly clear when giving instructions for a preparation. Here is an example; it is taken from his 'Book of Properties' and describes the preparation of white lead:

Take a pound of litharge, powder it well and heat it gently with four pounds of wine vinegar until the latter is reduced to half its original volume. Then take a pound of soda and heat it with four pounds of fresh water until the volume of the latter is halved. Filter the two solutions until they are quite clear and then gradually add the solution of soda to that of the litharge. A white substance is formed which settles to the bottom. Pour off the supernatant water and leave the residue to dry. It will become a salt as white as snow.

Of alchemical equipment, both substances and apparatus, Jabir gives a less systematic description than the later Muslim alchemist Razi (p. 88), but he does classify minerals into three groups. These are (*a*) spirits, or substances that volatilize completely on heating; (*b*) metals, or fusible substances that are malleable, sonorous, and possess a lustre; and (*c*) substances that, whether fusible or not, are not malleable and may be powdered. The spirits numbered sulphur, arsenic (i.e. arsenic sulphides, realgar and orpiment), mercury, camphor, and sal ammoniac. This is one of the earliest Arabic mentions of sal ammoniac, which for a time was imported from inner Asia; it was probably obtained there as a sublimate from burning coal-seams. Jabir, however, knew how to prepare it from organic matter, and distinguishes between the mineral form and the sal ammoniac 'from hair'.

Jabir recognized seven metals, namely gold, silver, lead, tin, copper, iron, and *khar sini*. The last of these has not been identified with certainty; the name signifies 'Chinese iron', and Muslim writers say that it was used in China to make mirrors with the power of curing ophthalmic maladies of sufferers looking into them. It was also cast into bells of a particularly melodious tone. Mme. Herrmann-Gurfinkal has suggested that it might have been arsenic, but according to B. Laufer it was an alloy chiefly composed of copper, zinc, and nickel; it had a silvery surface when polished and was known as *pai-t'ung* or white copper. There could have been very little of it available in Islam, and for the most part it was included among the metals only to bring their number up to seven when, as with Jabir, mercury was classed as a spirit.

Jabir's classification of minerals other than spirits and metals was not consistent, but in one book he sorts them into eight groups, according to whether they are (*a*) stony or not stony, (*b*) pulverizable or not pulverizable, and (*c*) fusible or non-fusible. Though such a system cannot be pressed very far, it is at least on sensible lines.

It is a matter of considerable interest that the Arabic text of the celebrated 'Emerald Table' of 'Hermes' was first discovered, by the present writer, in Jabir's 'Second Book of the Element of the Foundation'; before this discovery (1923) it had been known only in medieval Latin. The Emerald Table is discussed at greater length later in this book (p. 97). Also to be discussed later (p. 134) are the Latin works ascribed to Jabir or Geber.

Though Jabir's main alchemical preoccupation was that of all alchemists, namely to convert inferior metals into gold, he did not neglect to record chemical observations that seemed to him interesting or useful. He gives, in his book 'The Chest of Wisdom', the earliest known recipe for the preparation of nitric acid, and in another book he mentions the blue or green colour imparted to a flame by copper compounds. He describes processes for the preparation of steel and the refinement of other metals, for dyeing cloth or leather, for making varnishes to waterproof cloth and protect iron from rusting, for mordanting fabrics with alum, and for making an illuminating ink from 'golden' marcasite to replace the much more expensive one made from gold itself. He mentions the uses of manganese dioxide in glass-making, and he knew how to concentrate acetic acid by the distillation of vinegar. In various places he describes in some detail such typical chemical operations as calcination, crystallization, solution, sublimation, and reduction, going beyond the average alchemist by attempting to understand the changes that occur in these processes.

It would be unsatisfactory to end this account of 'Jabir' without expressing some opinion on the vexed question as to how much the *corpus* as a whole represents the work of one great mind. From the evidence described earlier, it would seem entirely unjustifiable to deny the existence of a historical alchemist named Jabir ibn Hayyan who flourished in the eighth century. It would be equally unjustifiable to minimize the great accretions made to original works by the Isma'ilites – the 'Isma'ilite jungle' as Stapleton puts it. But it is also clear that the Isma'ilites must have thought highly of the books they chose to interpolate and exrapolate so freely, and perhaps it would not be far from the truth to assume that the great bulk of the practical chemistry in the *corpus* represents the chemical knowledge of Jabir himself. How much of the theory and philosophizing was due to him and how much to the Isma'ilites can never be settled, but here again it seems likely that such theories as that of the 'balance' may be original. However all this may be, the strong impression is left that Jabir ibn Hayyan, a druggist's son who rose to be *persona grata* at the court of Harun al-Rashid, was a man of exceptional intellectual calibre, and we may echo the verse he wrote about himself:

My wealth let sons and brethren part;
Some things they cannot share –
My work well done, my noble heart,
These are mine own to wear.

THE TURBA PHILOSOPHORUM

We read in one of Jabir's books that several of the ancient philosophers, including Hermes, Pythagoras, Socrates, Aristotle, and Democritus, held an assembly to discuss the problems of alchemy. This is possibly the first reference to a celebrated alchemical work entitled *Turba Philosophorum* or 'Convention of Philosophers', the origin of which has puzzled scholars for several centuries. The *Turba* first appears in Latin manuscripts of the thirteenth century, and the earliest printed edition was published at Basel in 1572. It takes the form of a debate between a large number of philosophers and was held in great respect by generation after generation of alchemists.

The Latin version shows unmistakable signs of having been translated from the Arabic, and the content of the speeches make it equally clear that at least some of the material must have been derived from the Greek. The problems presented by the *Turba* attracted the attention of many historians of alchemy, and in 1931 Ruska published a monograph in which he definitely proved its Arabic origin and tried to fix its date by comparing it with other Arabic works. On this point, however, he could come to no definite decision, hesitating between the ninth, tenth, and eleventh centuries. He suggested that the *Turba* was an attack on the Greek alchemists and aimed at the liberation of alchemy from the plague of pseudonyms, and at basing it on a universally recognized natural philosophy. This particular suggestion was questioned by the present writer, but Ruska's proof that the *Turba* was Arabic in origin was fully confirmed in 1933, when Stapleton was able to show that a work of the tenth-century alchemist Ibn Umail (p. 102), contained passages excerpted from it.

There the matter rested until 1954, when an entirely new light was thrown upon it by Martin Plessner in what must be regarded as one of the most brilliant contributions to alchemical history made in the last half-century. Plessner first made the point that analysis of the *Turba* proves beyond doubt the unity of the work; therefore any Arabic work which contains quotations from it or parallels to it must be considered more recent. Ibn Umail died about 960, hence the *Turba* could not have been written much later than about 900. But the *Turba* contains a reference to deadly poison contained in the body of a woman, and though the expression here veils an alchemical meaning Plessner sees in it an affiliation to the Hindu myth of the poison-maiden, who kills men by her embrace. This myth was introduced to Islamic literature through the Arabic translation of a 'Book of Poisons' attributed to the Indian author Kautilya, a translation known to have been made in the first half of the ninth century. About this time lived an alchemical author named Uthman ibn Suwaid, of Akhmim (Panopolis) in Egypt, and among the titles of books attributed to him is 'The Book of the Controversies and Conferences of Philosophers'. Plessner suggests that this title may be evidence that the book, not extant so far as is known, was in fact the *Turba* – a suggestion that again leads to a date of composition of about 900. Akhmim was a Christian town with a noteworthy scientific tradition, where a great many people knew Greek, Coptic, and Arabic; this would explain the astonishing familiarity shown by the author of the *Turba* with Greek cosmology, and the manner in which the foundations of alchemy are made to appear in a cosmological guise.

In the Latin text, nine philosophers take part in the preliminary discussion, with the names Iximidrus, Exumdrus, Anaxagoras, Pandulfus, Arisleus, Lucas, Locustor, Pitagoras, and Eximenus. 'Anaxagoras' and 'Pitagoras' seem to indicate that the remaining seven names are mistransliterations of Greek names, and by transcribing them back into Arabic characters Plessner was able to show that the list should read Anaximander, Anaximenes, Anaxagoras, Empedocles, Archelaus, Leucippus, Ecphantus, Pythagoras, and Xenophanes – thus solving an age-old mystery.

These nine philosophers are all pre-Socratic, and Plessner demonstrates that in their speeches in the *Turba* they are reciting theories that, from classical sources, they are known to have held.

Anaximander discusses the Non-Limited (*Apeiron*); Anaximenes treats of the Air; Anaxagoras presents the conceptions of *Pietas* and *Ratio* as primary entities; Empedocles discusses the double function of the Air in separating Water and Earth and in mediating between Water and Fire; Archelaus treats of the Earth, the most compact, and of Fire, the finest element, as ruling the Universe; Leucippus talks of the Elements, without giving details, but referring apparently to the Full and the Empty as outlined by Diogenes Laertius; Ecphantus discusses the difference between the Upper and the Lower World, describing the former as containing beings composed of the two rare elements only; Pythagoras speaks of the simultaniety of all four Elements, which, according to him, are all primeval, and out of which all beings are composed; he does not, however, conceive of the presence of the four simultaneously in each being, but holds that the angels are composed of one Element only, the sun, moon and stars of two, and plants and animals of three, whereas only the human being is composed of all four Elements. Xenophanes, finally, postulates the coexistence of all four Elements, in varying mixtures, in all the beings of the world.

Plessner says that even where these opinions seem to contradict the doctrines of the Pre-Socratics as they are generally known, it is always possible to show items in the Greek tradition from which the tenets reported above were developed. The author did not misunderstand his sources, but in view of his purpose to connect cosmological teachings with alchemical matters he allowed himself to be somewhat tendentious.

The alchemical matters thus interwoven with the cosmological teachings are as follows: Anaximander praises the Air as a protector against Combustion [!]; Anaximenes points out the Dilution and Condensation of the Air, according to the various degrees of heat; Anaxagoras treats of the Density of matter, that increases from above to below; Empedocles speaks of the alchemical symbol of the Egg; Archelaus states the connexion between Fire and Earth; Leucippus presents the metaphor of Birth and Death commonly used in alchemy; Ecphantus outlines the alchemical doctrine of two Pairs of Elements; Pythagoras treats of the relations between Numbers and of the alchemical symbol of the Man; Xenophanes speaks of the ἓν τὸ πᾶν [*en to pan*, One is All], of the Putrefaction, and of the necessity of all four Elements being together.

It is in the closing speech of Xenophanes that the aim of the author becomes fully evident. This aim is to establish three theses, namely that the creator of the world is Allah, the God of Islam; that the world is of a uniform nature; and that all creatures of the upper as well as of the lower world are composed of all four elements. The preliminary discussion ends at this point, and the sixty-three further speeches constituting the remainder of the *Turba* are purely alchemical.

Other interesting relevant facts discovered by Plessner are, first, that all nine of the Pre-Socratics mentioned above appear in a book by Hippolytus (*c.* A.D. 222), one of the early Fathers of the Church, entitled 'Refutation of All Heresies', and that there is a close textual connexion between this book and the *Turba*. Secondly, in a book by the Greek alchemist Olympiodorus (*c.* 400), parallels are drawn between the doctrines of great alchemists and those of the philosophers, with precisely the same object of connecting cosmological with alchemical theory.

In summing up, Plessner says:

It is the threefold result of the cosmological discussion – the Koranic Creator-God, the Unified World, the Doctrine of the Four Elements – that gives the discussion its clear direction toward the chief subject of the *Turba*, alchemy. At the same time, alchemy is placed within the Islamic world of thought. In pursuing this purpose, the author displays a sovereign mastery of the doxographic literature * and an uncommon literary skill. He succeeds in producing a text which adds some genuinely new material to the doxography of the Pre-Socratics and represents the oldest evidence hitherto known of the penetration of the doxographic tradition into Islamic literature.

Plessner has thus gone very far to solve one of the most difficult conundrums of alchemy, and his further work on the subject is awaited with keen expectancy. It would be particularly interesting to know whether the *Turba* was originally written in Arabic, or whether it was the work of a Greek author given a tendentious Muslim twist at some later period. The latter is perhaps more likely.

RAZI (RHAZES)

Later than Jabir, though possibly contemporaneous with the writing of some of the Jabirian *corpus*, was Abu Bakr Muhammad ibn Zakariyya, known as Al-Razi, 'the man of Ray' (ancient *Rhagae*), from his birthplace, near Teheran. His lifetime extended from 825 or 826 to 925. In those days Ray was an important centre of culture, and Razi seems to have taken full advantage of the intellectual opportunities it offered. He studied philosophy, logic, metaphysics, and poetry, and was particularly fond of music, on which he is said to have written an encyclopedia; he himself was a skilful lute-player. The orientation of his interests towards medicine did not occur until he was thirty years old, when he paid a visit to Baghdad. There he is said to have met an apothecary with a wide knowledge of his subject and a memory crammed with details of unusual pathological experiences. Razi became so fascinated with the old man's stories that he made up his mind to embark upon a medical career, which he did with so much success that his books on the subject won him great fame not only in Islam but later, when they were translated into Latin, in western Europe: some of them were still prescribed for reading in universities in the Netherlands as late as the seventeenth century. One book, in which Razi for the first time clearly distinguished between smallpox

* Compilations of extracts from Greek philosophers.

and measles, has become a classic of medical literature; he also made important contributions to gynaecology and ophthalmology, and followed up his encyclopedia on music with another on medicine.

When the city hospital at Baghdad was to be rebuilt on a grand scale Razi is said to have been consulted about the plans and general lay-out, but he then returned to Ray and was placed in charge of the hospital there; however, he was not allowed to stay there very long and was recalled to Baghdad as physician-in-charge of the famous hospital he had helped to plan. When he retired he settled again in his native town.

Razi excelled not only as a practitioner and medical author but as a teacher. One of his biographers has drawn a graphic picture of Razi as an old man, seated on the paving of the courtyard of the hospital surrounded by his pupils. The advanced students sat in a ring nearest him, and in outer concentric rings sat those whom we might call the second-year and first-year men. Razi, the fount of wisdom at the centre, expounded to his immediate *entourage*, and the information, suitably simplified, was passed on to the less experienced. Razi is said to have been of a kindly and courteous disposition, never too busy to receive visitors and always ready to help the suffering; in spite of this he was such a prolific author that his industry must have been enormous – we are, indeed, told that he was invariably at his desk when his other duties allowed. It is not surprising that in the end he developed eye-trouble resulting in blindness, though detractors put this infirmity down to Razi's inordinate consumption of beans.

Like the majority of physicians of medieval times, Razi was led to the study of alchemy and is said to have written a score of books on the subject; they have not all survived, but one of them, 'The Book of the Secret of Secrets', has been translated into German and furnished with a commentary by Ruska. Much of his alchemical work has also been studied by Stapleton, who places him on an intellectual level with Galileo and Boyle. Like Jabir – or like the author or authors of much of the Jabirian *corpus* – Razi showed the spirit of free inquiry characteristic at that time of the Isma'ilites, and adopted much of their philosophical background. His theory of matter has been thus described by G. Heym:

> He taught that there were five eternal principles: the Creator, the soul, matter, time, and space. Bodies were composed of indivisible elements and of empty space that lay between them. These atoms or elements were eternal and possessed a certain size. The characteristics of the four elements earth, water, air, and fire, that is, their lightness and their heaviness, their transparency and colour, their softness and hardness, these characteristics were determined by the density of the elements, in other words by the measure of the emptiness of the spaces between them. These spaces determined the natural motion of the elements: water and earth moved downwards, air and fire upwards.

Though Razi did not accept Jabir's elaborate theory of the 'balance' he nevertheless believed that, basically, all substances were composed of the four 'elements', and that therefore the transmutation of metals was possible; the object of alchemy was to effect this transmutation of base metals into silver or gold by means of elixirs, and also to 'improve' valueless stones such as quartz and even glass with similar elixirs and so convert them into emeralds, rubies, sapphires, and the like. Razi followed Jabir in assuming that the proximate constituents of metals were mercury and sulphur (or an inflammable oil), but sometimes suggests a third constituent of a salty nature – an idea which occurs very frequently in later alchemical literature (p. 174). As to the elixirs, which strangely enough Razi never refers to as the philosophers' stone, they were of varying powers, ranging from those which could convert only one hundred times their own weight of base metal into gold to those that were efficient on 20,000 times.

A study of Razi's writings, however, especially 'The Book of the Secret of Secrets', conveys the impression that he was much more interested in practical chemistry than in theoretical alchemy. The 'Secret of Secrets' foreshadows a laboratory manual, and though the procedures described are often difficult to interpret they are probably representative of experiments that Razi himself had carried out. Razi in fact brought about a revolution in alchemy by reversing the relative importance of experiment and speculation; whereas earlier adepts had swamped practice in floods of unsupported hypothesis, Razi felt that if success were to be obtained it would be from work in the laboratory and not from lucubrations in the study.

From the lists he gives of materials and apparatus it is evident that his own laboratory was very well equipped. It had beakers, flasks, phials, basins, glass crystallizing dishes, jugs, casseroles, candle-lamps, naphtha-lamps, braziers, furnaces called athanors, smelting-furnaces, files, spatulas, hammers, ladles, shears, tongs, sand-baths, water-baths, filters of hair-cloth and linen, alembics, aludels, funnels, cucurbits, and pestles and mortars. In addition, Razi gives details of the construction of more complicated pieces of apparatus from these and other units.

His store-cupboard contained not only specimens of all metals then known, but pyrites, malachite, lapis lazuli, gypsum, haematite, turquoise, galena, stibnite, alum, green vitriol, natron, borax, common salt, lime, potash, cinnabar, white lead, red lead, litharge, ferric oxide, cupric oxide, verdigris, and vinegar. Stapleton has adduced reasons for believing that Razi was also familiar with caustic soda and glycerol. It is uncertain whether he was acquainted with sulphuric and nitric acids, but, since nitric acid was known to Jabir, he probably was.

Razi's systematic and orderly mind led him to draw up a scheme for the classification of substances used in alchemy; here, for the first time, we meet with the division of substances – so familiar from the nursery and later from the B.B.C. – into animal, vegetable, and mineral. Razi's scheme is as shown on page 93.

The chemical processes described or mentioned by Razi include distillation, calcination, solution, evaporation, crystallization, sublimation, filtration, amalgamation, and ceration, the last named being a process for converting substances to pasty or fusible solids. Most of these operations were pressed into service for attempts at transmutation, which according to Razi were generally conducted as follows. First, the substances to be employed must be purified by distillation, calcination, amalgamation, or other appropriate treatment. Having freed the crude materials from their impurities, the next step was to reduce them to an easily fusible condition by means of ceration, which should result in a product that readily melted, without any evolution of fumes, when dropped upon a heated metal plate. After ceration, the product was to be further disintegrated by the process of solution, which included dissolving in 'sharp waters'; these were generally not acid liquids but alkaline and ammoniacal, though lemon juice and sour milk, which are weakly acidic, were sometimes employed.

The solutions of the various substances, suitably chosen in proportion to the amount of 'bodies', 'spirits', etc., they were supposed to possess, were then brought together. The combined solutions were finally subjected to the process of coagulation or solidification, and if the experiment was successful the substance resulting would be the – or an – elixir. Alchemists were always optimistic, and if the outcome of such a laborious proceeding as that just described failed to transmute base metal into gold it might often be credited with therapeutic value – and not invariably without justification. Practical alchemy had in fact much to offer medicine and, as we shall see later, gave rise in due course to a period of medical chemistry or iatrochemistry (Chapter 8), the forerunner of our modern chemotherapy.

There is no doubt that Razi's emphasis upon practical research led to the advances in pharmacology well exemplified in the work of a late tenth-century Persian physician named Abu Mansur Muwaffak, in whose writings many important chemical facts are apparently recorded for the first time. Thus Abu Mansur was probably the first to make a clear distinction between sodium carbonate (soda) and potassium carbonate (potash), which are in many respects very similar. He describes arsenious oxide (white arsenic) as a pure white powder, and mentions another white substance, now known as silicic acid, obtainable from the bamboo. He says that when gypsum is heated it yields a sort of lime which, mixed with white of egg, forms a plaster of great service in the treatment of fractured bones: the 'lime' was in fact plaster of Paris. He observes that though antimony is a dark solid, a freshly cut surface of it shows a fine metallic lustre; that on exposure to air copper is converted into a greenish mass similar to malachite, but if strongly heated in air it yields

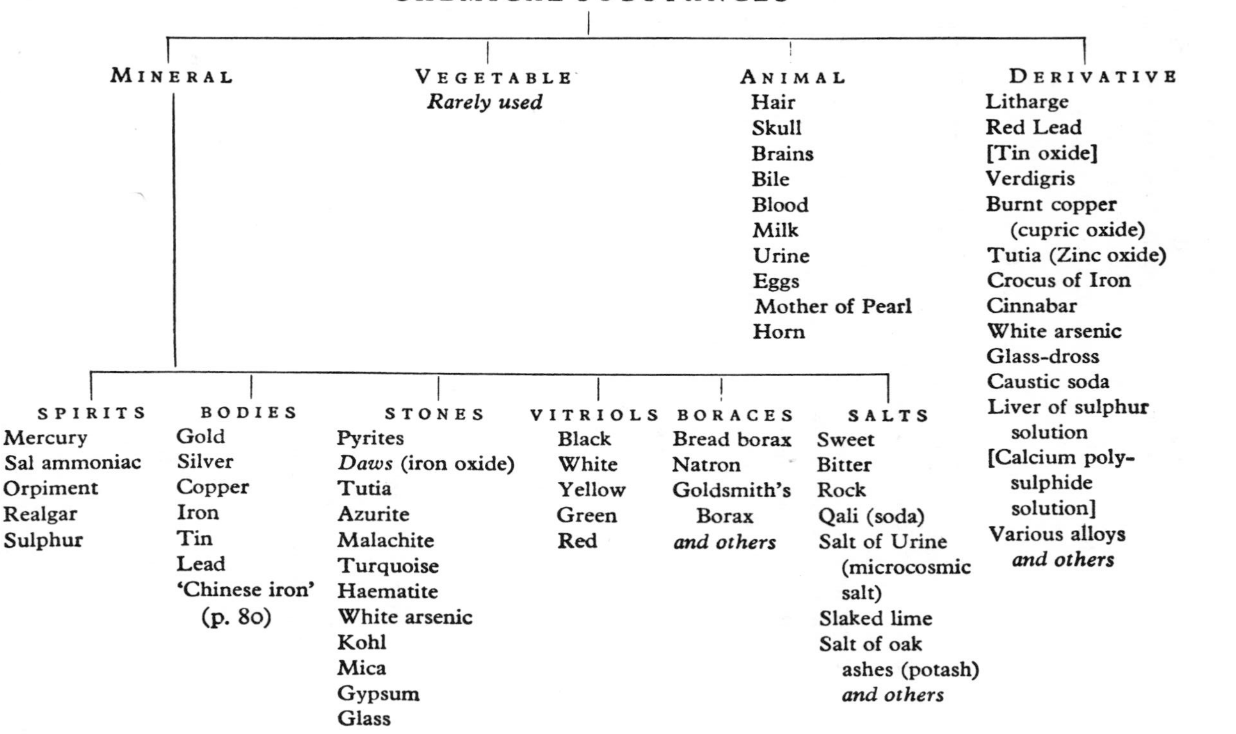
CHEMICAL SUBSTANCES
MINERAL
VEGETABLE
Rarely used
ANIMAL
Hair
Skull
Brains
Bile
Blood
Milk
Urine
Eggs
Mother of Pearl
Horn
DERIVATIVE
Litharge
Red Lead
[Tin oxide]
Verdigris
Burnt copper (cupric oxide)
Tutia (Zinc oxide)
Crocus of Iron
Cinnabar
White arsenic
Glass-dross
Caustic soda
Liver of sulphur solution
[Calcium poly-sulphide solution]
Various alloys
and others
SPIRITS
Mercury
Sal ammoniac
Orpiment
Realgar
Sulphur
BODIES
Gold
Silver
Copper
Iron
Tin
Lead
'Chinese iron' (p. 80)
STONES
Pyrites
Daws (iron oxide)
Tutia
Azurite
Malachite
Turquoise
Haematite
White arsenic
Kohl
Mica
Gypsum
Glass
VITRIOLS
Black
White
Yellow
Green
Red
BORACES
Bread borax
Natron
Goldsmith's Borax
and others
SALTS
Sweet
Bitter
Rock
Qali (soda)
Salt of Urine (microcosmic salt)
Slaked lime
Salt of oak ashes (potash)
and others

a black substance (copper oxide) that may be used to darken the hair. Such accurately observed facts indicate that a by-product of alchemy was a steadily increasing body of reliable chemical knowledge, a trend which Razi did most to establish and for which he deserves the gratitude of succeeding generations.

AVICENNA

The state of the Muslim empire at the end of the tenth century [says Baron Carra de Vaux] may be represented by that of an undisciplined and stormy feudalism, where under an enervated and disorganized central authority a crowd of vassal powers spring up one after another, dominate a part of the empire and are then eclipsed. Races and creeds come into conflict, advancing or retreating according to the fortunes of the political adventurers who represent them. In general, the Arab spirit is in decline; the old Persian spirit awakes from time to time, but never quite succeeds in freeing itself from the chaos, hindered as it is by outbursts of barbarism due chiefly to the Turkish element. Nevertheless, science pursues its destinies, in the shelter of the ephemeral protection afforded it here and there by a few princely personages. It is in such circumstances, whose troubled and tempestuous character is reflected in his life, that Avicenna for the first time gave a clear, ordered, and complete expression to that calm and grandiose system that we call scholasticism.

Abu Ali ibn Sina, known in Europe as Avicenna, held views on alchemical claims that cannot be omitted from any history of alchemy. This Islamic genius, the 'Prince of Physicians', who has been described as the Aristotle of the Arabians and certainly the most extraordinary man the nation produced, was not in fact an Arab but a Persian. He was born at Afshana, near Bukhara, in 980, and his father was a native of Balkh. After the birth of Avicenna's younger brother the family moved into Bukhara itself, where a tutor was engaged to instruct the future philosopher in the Koran and in Arabic poetry. The boy made such rapid progress that additional tuition was soon required, and he was taught arithmetic by a greengrocer, law by an ascetic named Ibrahim, and Euclid and logic by a wandering scholar called Natili whom his father lodged in the house for his son's benefit. Natili seems to have had but a slender stock of knowledge, and Avicenna, having discovered this, applied himself with energy and resolution to a course of hard private study. Among many other subjects he studied medicine, 'which', he says, 'is not difficult', and by the age of sixteen he had advanced so far that adult qualified physicians came to learn new methods of treatment from him.

Appointed physician to one of the princes of the country at the tender age of seventeen, Avicenna held many important posts in after years, on one occasion being grand vizier or prime minister to Shams al-Daula at Hamadhan. He later went to Ispahan, and after an eventful life died at Hamadhan in 1036 or 1037. In his comparatively brief span of existence he accomplished an amazing mass of literary, medical, philosophical, and scientific work, and became an almost legendary hero to his co-religionists and even to medieval Europe. Altogether he wrote over a hundred books, and though some of them are quite short his celebrated 'Canon of Medicine' contains over a million words.

This is not the place to describe Avicenna's contributions to medicine, but it may be mentioned in passing that he had original ideas on psychiatry and nervous affections, realized that phthisis was contagious, and believed that certain diseases could be transmitted by soil and water. In the pharmacological section of the 'Canon' he mentions no fewer than 760 drugs, including the narcotics mandragora, opium, hemlock, and cannabis. For several hundred years, Avicenna shared with Razi the position of being in medicine an authority from whom there was no appeal.

Like Razi again, Avicenna was deeply interested in music, and his studies of musical theory were far ahead of those current in Europe at that time. In physics, he treated of heat, energy, gravity, and motion, and suggested that light travelled with a finite velocity. He had a considerable knowledge of mathematics from the philosophical point of view, invented a kind of vernier, and made astronomical observations.

What did such a versatile and acute mind think about alchemy? His contemporaries were almost unanimous in believing that transmutation of the metals was possible, and many of them held that it had been successfully accomplished; others – a minority – while holding it to be theoretically irrefutable were sceptical about individual claims to success. Avicenna did in fact express himself very clearly on the matter. His remarks are to be found in a section of his work known as the 'Book of the Remedy', which he wrote at Hamadhan about 1021–3. A medieval Latin translation of this section had long been known under the title *De Mineralibus*, 'On Metals', and was usually ascribed to Aristotle; in 1927, however, Holmyard and D. C. Mandeville showed that the *De Mineralibus* was partly a direct translation and partly a résumé of chapters in the 'Book of the Remedy' and thus represented Avicenna's own views.

The relevant portions of the work deal with the constitution of metals, and here Avicenna follows Jabir very closely. He says that the proximate constituents of metals are mercury and sulphur, or bodies closely resembling them. If the mercury is pure, and if it is commingled with the virtue of a white sulphur that neither induces combustion nor is impure, but on the contrary is more excellent than that prepared by the alchemists, then the product is silver. If the sulphur, besides being pure, is even better than that just described, and whiter, and if in addition it possesses a tinctorial, fiery, subtle, and non-combustive virtue, it will solidify the mercury into gold. Then again, if the mercury is of good substance, but the sulphur that solidifies it is impure, possessing a property of combustibility, the product will be copper. If the mercury is corrupt, unclean, lacking in cohesion and earthy, and if the sulphur also is impure, the product will be iron. As for tin, it is probable that its mercury is good but that its sulphur is corrupt; and that the commingling of the two is not firm, but has taken place, so to speak, layer by layer, for which reason the metal 'shrieks'. This is an interesting reference to, and an attempted explanation of, the well known 'cry of tin', which modern chemistry ascribes to friction of the crystalline particles. Lead, says Avicenna, is probably formed from an impure, fetid, and feeble sulphur, for which reason its solidification has not been thorough.

One might have expected that, adhering in the main to what were generally accepted alchemical views of metallic structure, Avicenna would have given credence to the practicability of converting one metal into another. On the contrary, he expresses complete incredulity concerning transmutation. There is little doubt, he says, that the alchemists can contrive to make solids in which the qualities of metals are perceptible to the senses, though the alchemical substances are not identical in principle or in perfection with the natural ones, but merely bear a resemblance and relationship to them. As to the claims made by the alchemists, he says that it must be clearly understood that it is not in their power to bring about any true change of the metallic species. They can, however, produce excellent imitations, whitening a red metal so that it closely resembles silver, or tinting it yellow so that it closely resembles gold. They can also colour a white metal in such a way as to make it resemble gold or copper, and they can free lead and tin from most of their defects and impurities. Yet in metals so treated the essential nature remains unchanged; they are merely so dominated by induced qualities that errors may be made concerning their real nature. 'I do not deny,' he proceeds, 'that such a degree of accuracy in imitation may be reached as to deceive even the shrewdest, but the possibility of transmutation has never been clear to me. Indeed, I regard it as impossible, since there is no way of splitting up one metallic combination into another. Those properties that are perceived by the senses are probably not the differences which distinguish one metallic species from another, but rather accidents or consequences, the essential specific differences being unknown. And if a thing is unknown, how is it possible for any one to endeavour to produce it or to destroy it?' It is very clear that Avicenna was contemptuous of the pretensions of alchemy, for he winds up this passage by remarking that there was much he might have said on the subject, but that it would probably have been a sheer waste of time – a remark that the Latin translator tactfully omitted from his version.

As a matter of fact, scepticism concerning alleged transmutations had long existed. Some denied the possibility altogether; others agreed that transmutation might be effected, but only by natural or 'white' magic. Even in Jabir's time disbelief had been expressed, and it is said that Razi himself wrote a book to confound the sceptics, among whom was the celebrated scholar and translator Hunain ibn Ishaq (p. 67). Avicenna's attack did not go unanswered. His arguments were examined carefully by the vizier Al-Tughra'i, better known as an accomplished poet than as an alchemist, and were shown to be inconsistent with views that Avicenna had himself expressed in other passages in the same book; and so the controversy went on. Avicenna's opinions had little support; here, as in many other branches of learning, he was in advance of his time, and the alchemists pursued their search for elixirs undisturbed by doubts and with undiminished enthusiasm.

Throughout the history of alchemy there were always rogues and charlatans whose trickery brought honest attempts at transmutation into disrepute. One such knave about this time arrived at Damascus with some gold filings which he mixed into a paste with charcoal, various drugs, flour, and fish-glue. He then rolled the paste into small pellets which he allowed to dry. Clothing himself as a dervish he took the pellets to a druggist and sold them for a few pence under the name of 'tabarmaq of Khorassan'. Next he assumed a rich cloak, engaged a servant, and went to the mosque, where he scraped acquaintance with several notable persons. He told them that he was an expert alchemist and could make untold wealth in a single day, a boast that soon came to the ears of the vizier, who ordered his presence at the court. The Sultan expressed his desire to witness a transmutation, which the charlatan readily agreed to demonstrate if he could be provided with the requisite chemicals. The recipe he produced included a certain amount of tabarmaq of Khorassan, and while all the rest of the drugs were easily obtained no trace of tabarmaq could at first be discovered. The charlatan insisted, however, that tabarmaq was essential, and when the druggists' shops had been well searched the discovery was at length made – of course in the shop of the druggist to whom the tabarmaq had been sold earlier, and who said that he had obtained it from a dervish.

The pellets were bought, and the charlatan ordered the ingredients to be placed in a crucible and strongly heated. When all was sufficiently hot, 'Take out the crucible', he said. It was taken out, cooled, and turned upside down, when a fine lump of gold rolled out.

The Sultan, deeply impressed by the success of the experiment, ordered the self-styled alchemist to be rewarded, and the next step was now to find a further supply of tabarmaq. Search failed to reveal even a single further pellet in Damascus, but the alchemist said he knew of a cavern in Khorassan where large supplies were to be found. He suggested to the Sultan that an expedition should be sent to bring back a goodly quantity of the rare substance. But the Sultan – no doubt reacting just as the alchemist had foreseen – was unwilling that the provenance of tabarmaq should become known to anyone else, and commanded the alchemist to go alone. After a show of reluctance the man accepted, and was furnished by the Sultan with everything needed for the journey: a tent, a travelling kitchen, sugar, carpets, stuffs and silks, manufactured objects from Alexandria, and, in addition, a large sum of money. Thus equipped, he set out – and that was the last that was seen of him.

THE EMERALD TABLE

Before considering some of the later Muslim alchemists we may turn aside for an account of the celebrated 'Emerald Table' or *Tabula Smaragdina*, since the earlier known versions of this expression of alchemical dogma are in Arabic. The Table itself is ascribed to Hermes or the Egyptian Thoth, god of mathematics and science, and it has been well known since the early Middle Ages in a Latin rendering. It purports to summarize the principles of change in Nature and therefore lies at root of alchemical doctrine. An English version, by R. Steele and (Mrs) D. W. Singer, runs as follows:

> True it is, without falsehood, certain and most true. That which is above is like to that which is below, and that which is below is like to that which is above, to accomplish the miracles of one thing.
>
> And as all things were by the contemplation of one, so all things arose from this one thing by a single act of adaptation.
>
> The father thereof is the Sun, the mother the Moon.
>
> The Wind carried it in its womb, the Earth is the nurse thereof.

It is the father of all works of wonder throughout the whole world.

The power thereof is perfect.

If it be cast on to Earth, it will separate the element of Earth from that of Fire, the subtle from the gross.

With great sagacity it doth ascend gently from Earth to Heaven.

Again it doth descend to Earth, and uniteth in itself the force from things superior and things inferior.

Thus thou wilt possess the glory of the brightness of the whole world, and all obscurity will fly far from thee.

This thing is the strong fortitude of all strength, for it overcometh every subtle thing and doth penetrate every solid substance.

Thus was this world created.

Hence there will be marvellous adaptations achieved, of which the manner is this.

For this reason I am called Hermes Trismegistus, because I hold three parts of the wisdom of the whole world.

That which I had to say about the operation of Sol is completed.

Steele and Singer say: 'It is no part of our intention to give a definite meaning to these cryptic utterances' – a decision to which we may well subscribe. At the same time the Table does vaguely indicate the alchemical belief that there is a correspondence or interaction between celestial and terrestrial affairs, and that all the manifold forms in which matter occurs have but a single origin. A universal soul or spirit permeates both macrocosm and microcosm, and this unity in diversity implies the possibility of transmutation. The Sun and Moon in the Table may represent gold and silver, as they usually do; but later alchemists seem sometimes to have regarded them in this context as referring to sulphur and mercury. Alchemical commentators through the centuries attached their own interpretations to the dicta comprised in the Table, and in so far as any agreement can be found between them it is in the general idea that the powers of the cosmic soul must somehow be concentrated in a solid, the philosophers' stone or elixir, which would then be able to carry out the transmutations that the alchemists desired.

Whatever its meaning, there is no doubt that the Emerald Table is one of the oldest and most long-lived of all alchemical documents. In its original form it was alleged to have been found in a cave, inscribed on a plate of emerald held in the hands of the corpse of thrice-greatest Hermes, Hermes Trismegistus. Embellishing details of the legend are that the writing was in Phoenician characters, and that the fortunate discoverer was Sara, the wife of Abraham, who chanced to enter a cave near Hebron some long time after the Flood. Other versions ascribe the discovery to Alexander the Great and to Apollonius of Tyana, who flourished in the first century A.D. From the clouds of myth surrounding the early history of the Table it is not possible to discern the original authorship, but during the last few years a good many new facts about it have come to light.

As already mentioned (p. 81), an abridged Arabic text of the Tabula was discovered by the present writer in 1923, in one of the books of the Jabirian *corpus*. Shortly afterwards, another Arabic version was discovered by Ruska in a book called 'The Secret of Creation', ascribed to Apollonius. Jabir himself, when giving the *Tabula*, says that he is quoting from Apollonius. Now Kraus was able to show that 'The Secret of Creation' was written, at least in its final form, during the caliphate of Al-Ma'mun (813–33), and that it shows parallels with a book written during the same period by Job of Edessa, a scholar whose translations from Syriac into Arabic won the praise of so severe a critic as Hunain ibn Ishaq. It seems likely, therefore, that even if Job did not write 'The Secret of Creation', both he and its author were using the same earlier sources, one of which Kraus showed to be the writings of Nemesius, bishop of Emesa (Homs) in Syria in the second half of the fourth century. Nemesius wrote in Greek, but his book 'On the Nature of Man' does not contain the *Tabula*. From these data one may conclude that the oldest known form of the Table, namely that in Arabic, was probably translated from Syriac but that it may ultimately have been based on a Greek original. Whether it went back as far as Apollonius himself is a question to which no answer is possible; he was a Neo-Pythagorean and wonder-worker and it would have been natural for him to be interested in alchemy. In any case it seems likely that the *Tabula* came to Islam from Syria rather than from Alexandria, for Arabic accounts of its discovery usually mention the Flood – Noah took it with him in the Ark – and the Flood was unknown in Egypt.

An Arabic author says that there were three persons called Hermes. The first of them, Hermes Trismegistus, lived before the Flood and was the grandson of Adam. He wrote many books about the knowledge of celestial and terrestrial things and built the pyramids of Egypt. He was the first patron of science and mathematics, and the first to wear sewn clothes. Hermes the second was an inhabitant of Babylon and lived after the Flood. He excelled at science, medicine, philosophy, and mathematics, and was the teacher of Pythagoras. Hermes the third lived in Egypt, was a physician and philosopher, wrote a book on poisonous animals, another on alchemy and the chemical arts, and was also an expert town-planner.

Such myths, which grew and became more circumstantial as time went on, were accepted by alchemists as explaining the origin of alchemy, and account for the frequency with which it was known as the Hermetic Art – as also for the modern chemist's 'hermetically sealed'. It would take us much too far afield to discuss in detail the influence that the body of Hermetic literature had on Muslim alchemy, but it was certainly responsible to a large extent for very numerous religious, mystical, magical, and astrological tenets intermingling – often inextricably – with alchemical doctrine in the narrower sense. Particularly it emphasized the influence that the heavenly bodies were supposed to exert not only on man but on metals and other objects.

LATER MUSLIM ALCHEMISTS

Under the caliphate of Al-Hakam II, who ruled in Spain from 961 till 976, there flourished a brilliant group of Spanish-Arab scholars, among them being Maslama ibn Ahmad; though a native of Cordova he is usually known as Al-Majriti or 'of Madrid' because of his long residence there. He was educated partly in the East, and while there seems to have come into contact with the celebrated Encyclopedists of Islam, the 'Brethren of Purity', whose 'Letters', which cover a wide range of contemporary knowledge, he is said by some authorities to have brought back with him to Europe in a new recension. He is known chiefly for his astronomical work, which included the revision of some Persian astronomical tables into Arabic chronology, a commentary on the *Planisphaerium* of Ptolemy, and a treatise on the astrolabe; both the latter works were early translated into Latin and had what was for those days a wide circulation. To him are ascribed an important alchemical treatise called 'The Sage's Step' and a magical work called 'The Aim of the Wise'. The magical work was translated into Spanish in 1256, by command of Alfonso the Learned, King of Castile and León from 1252 to 1284, and later a Latin version became very popular under the name of 'Picatrix' (a corruption of Hippocrates via Bugratis). There is a reference to it in *Pantagruel*, where Rabelais (1495–1553) speaks of *'le reverend père en Diable Picratis, recteur de la faculté diabolologique'* at Toledo. 'The Sage's Step' does not appear to have been translated, but an analysis of it was published by the present writer in 1924. It is there shown that its ascription to Al-Majriti must be false, since from internal evidence it could not have been written till after 1009 while Al-Majriti died in 1007; however, it is possible that in this case, as in so many others, the work was edited and perhaps enlarged after the original author's death. The same remarks apply to the authorship of 'Picatrix'.

'The Sage's Step' is of interest for several reasons. First, the author expresses his views as to the preliminary training necessary for the alchemical aspirant. This should include the study of mathematics in the pages of Euclid and Ptolemy, and of the natural sciences as taught by Aristotle or Apollonius of Tyana. Next, the student should practise his hand in operation, his eye in examination, and his mind in reflection over chemical substances and reactions. Since Nature's behaviour is invariable, for she never does the same thing in different ways, the chemist must strive to follow Nature, whose servant indeed he is, like the physician. The latter diagnoses the disease and administers a remedy, but it is Nature that acts.

Secondly, 'The Sage's Step' contains very precise and intelligible instructions for the purification of gold and silver by cupellation and in other ways, serving to show that contemporary alchemy was not all head-in-air but knew the discipline of the laboratory. And thirdly, the author of the book describes an experiment, on the preparation of what is now called mercuric oxide, carried out on a quantitative basis. Very seldom in alchemical literature do we find even the glimmering of an idea that pursuing the changes in weight that occur during a chemical reaction might lead to significant results: a procedure that, first methodically applied by Joseph Black in the middle of the eighteenth century, has been for two hundred years a guiding principle in the science of chemistry.

Contemporary with Maslama al-Majriti was an alchemist named Muhammad ibn Umail, about whose life, since he lived in seclusion, very little is known. However, some of his works have survived, notably a book entitled 'The Silvery Water and the Starry Earth', which is a commentary on his own alchemical ode 'Epistle of the Sun to the Crescent Moon'. Both of these works were translated into Latin in the Middle Ages, the former under the title *Tabula Chemica* and ascribed to 'Senior Zadith, son of Hamuel', and the latter as *Epistola Solis ad Lunam Crescentem.* The 'Silvery Water and Starry Earth' is valuable on account of its very numerous quotations from earlier alchemical authors, and particularly because it provides another example of the extent to which Muslim alchemy was accepting – and developing – Hermetic doctrine. Some of the 'Sayings of Hermes' quoted by Ibn Umail have been shown by Stapleton, Lewis, and Sherwood Taylor to have come from Greek originals; others are probably apocryphal and of tenth-century Arabic composition.

During the twelfth and thirteenth centuries alchemical theory and practice were transmitted to the West, a process to be described in the next chapter; and the transmission was accompanied by a decline in the number of noteworthy Arabic writers on the subject. Such books as were written were for the most part rearrangements or compendia of earlier ones or commentaries on them. Few authors of the time need mention here. One of them, Ibn Arfa Ras, enjoyed much fame among later Muslim adepts for his alchemical poem entitled 'Particles of Gold', which, whatever its merit in expounding alchemical theories, is at least of high literary achievement. A second writer of standing was the thirteenth-century Abu'l-Qasim al-Iraqi, whose book 'Knowledge Acquired concerning the Cultivation of Gold' was published, with an English translation and a commentary, by Holmyard in 1923. It gives a good picture of contemporary Islamic alchemical ideas, but is not altogether original – drawing, for instance, upon Ibn Umail's 'Silvery Water and Starry Earth' and other sources. It was, however, considered important enough by Aidamur al-Jildaki, who lived in the first half of the fourteenth century, to deserve a lengthy commentary, which he provided in his 'Book of the End of the Search'. Little is known of Jildaki's life, as of Ibn Umail's, but he was certainly not a recluse. He dwelt in Cairo for a time, then spent seventeen years in extensive travel to meet the principal alchemists of the time and to collect alchemical writings. Most of the remainder of his life he must have devoted to authorship, for he wrote at least twenty-five books, some of them very voluminous. Their value lies in the great number of quotations that Jildaki fortunately saw fit to incorporate in them, a value enhanced by the general accuracy with which the quotations are made. In many cases the original works from which the quotations were made are still in existence, and examination of them shows that Jildaki was a careful copyist; we may therefore with fair confidence accept as genuine other passages of which no earlier provenance is known.

The great bulk of Jildaki's work still awaits detailed study, but a full examination of the 'End of the Search' was made by M. Taslimi in 1954. The richness of the material offered by Jildaki may be judged from the fact that in this book alone he quotes from, or mentions, no fewer than forty-two works of Jabir and a large number of those of other authors, including Ibn Umail, Avicenna, Maslama al-Majriti, Khalid ibn Yazid, and Razi. Another of Jildaki's books that would well repay study is the 'Book of Proof', which contains a commentary on the 'Book of the Seven Idols' ascribed to Apollonius of Tyana. Taslimi says that there is a great deal of similarity between the ideas contained in the quotations from Jabir given in the 'End of the Search' and those found in the Latin works of Geber (p. 134);

he does not, however, think the correspondence sufficiently close to establish a definite affiliation.

Though after Jildaki there is no outstanding figure in Muslim alchemy, that does not mean that the Art ceased to be cultivated. There are, indeed, still industrious – and sometimes learned – alchemists in Islam. One of them, Al-Hajj Abdul-Muhyi Arab, who died only a few years ago, was a personal friend of the author's and often engaged him in discussions concerning Jabir, Khalid, and other Arabic authorities. He believed that Khalid's contributions to the Art had been undervalued, and placed him almost on the same level as Jabir. Abdul-Muhyi was a well-educated man, knew the Koran by heart, and had been lecturer on Koranic exegesis at the university of Lahore. As Mufti of the Shah Jehan Mosque at Woking, he spent practically the whole of his salary on conducting alchemical experiments on a portable blacksmith's forge, but, alas, without success. One of his last acts was to write a letter of introduction to an alchemist friend at Fez, the outcome of which was to give the author the privilege of being taken to see a subterranean alchemical laboratory in the old part of that city.

6

EARLY WESTERN ALCHEMY

TRANSMISSION OF ALCHEMY FROM ISLAM TO WESTERN EUROPE

WHILE early medieval Europe was by no means destitute of skilful dyers, painters, glass-makers, goldsmiths, metallurgists, and other craftsmen or technicians, there appears to have been no knowledge of alchemy in the West until it was introduced from Islam, a process beginning in the twelfth century. The few Latin manuscripts from before that time dealing with chemical matters are merely collections of practical recipes.

Until the twelfth century almost the sole contact between Islam and Christian Europe was through the crusades, which were clearly not favourable to the transmission of learning. Soon after 1100, however, European scholars began to discover that the Saracens were possessed of much knowledge and ancient wisdom, and the bolder spirits began to travel in Muslim lands in search of learning and enlightenment. Sicily, an appanage of Islam from 902 to 1091, was captured by the Normans in the latter year, and the island thus became a centre of diffusion of Arabic learning. It was, however, in Spain – still largely under Moorish control – that the greatest activity prevailed. Students were welcomed to the colleges and libraries at Toledo, Barcelona, Segovia, Pamplona, and other Spanish towns, and study was soon followed by translation.

One of the earliest of the translators was the Englishman Robert of Chester, who is believed to have been a native of Ketton in Rutland; his association with Chester perhaps indicates that he was educated at the well-known school in that city. In 1141 he and his friend Hermann the Dalmatian were living in Spain near the Ebro, studying the arts of alchemy and astrology. Here they were found by Peter the Venerable, Abbot of Cluny, who persuaded them to translate the Koran into Latin, a task which they immediately started, and which took them two years.

Robert then turned to the translation of an Arabic alchemical book, the 'Book of the Composition of Alchemy', completing it on 11 February 1144. This was the first work on alchemy to appear in Latin Europe, so we can date the introduction of the Art very precisely. In the preface to his translation Robert says, 'Since what Alchymia is, and what its composition is, your Latin world does not yet know, I will explain in the present book'. He continues,

> Alchymia is a bodily substance compounded of one, or by one, thing, and more precious by conjoining nearness and effect, and with the same natural commixion converting naturally with better policies. But in that which followeth, this which we have spoken shall be explained, where we will intreat at large of the composition of it. And although our wit be but raw and our Latin but little, yet we have taken in hand to translate this great work out of the Arabic tongue into Latin.... And it hath seemed good unto me to set my name in the beginning of the preface, lest any man should attribute this our labour unto himself and also challenge the praise and desert as due unto himself.

The chief point of interest about the 'Composition of Alchemy' is that it tells the story of Khalid ibn Yazid and his reputed master Morienus; thus the earliest figure of Muslim alchemy becomes also the earliest of European alchemy. The original legend is here expanded and provided with numerous details of an obviously legendary nature, but that Robert should have happened to select it for translation out of all the very many alchemical works available is a remarkable coincidence.

For some time, Robert was archdeacon of Pamplona, in northern Spain, but he returned to London in 1147, and again in 1150. In addition to his services to alchemy, he translated the 'Algebra' of the celebrated Muslim mathematician Al-Khwarizmi (from whose name comes our word algorism), and thus must be credited with introducing to Western Europe not only a new 'science' but a new branch of mathematics. He also calculated a set of astronomical tables for the meridian of London, wrote a treatise on the astrolabe, and was the first to use the word sine (Arabic *jaib*) in its accepted trigonometrical sense. The translation of an Arabic commentary on the Emerald Table (p. 97) is possibly attributable to him.

Another English scholar active in translation at this time was Adelard of Bath, the dates of whose birth and death are unknown but whose major work was accomplished between 1116 and 1142. He studied at Tours and Laon, then travelled widely in Italy, Sicily, Asia Minor, and Africa before returning to England. When he arrived home he was greeted by friends, of whom he inquired about the state of affairs in his native land. He was told that 'princes' were 'violent, prelates wine-bibbers, judges mercenary, patrons inconstant, the common men flatterers, promise-makers false, and everyone in general ambitious'. Adelard rejected the idea of conforming to such moral depravity, and as he could not do anything to prevent it he welcomed the suggestion of his nephew that he should rather describe what he had acquired in the way of new information from his Islamic studies.

He prefaces his remarks by saying that his contemporaries were prejudiced against fresh ideas, and that therefore what he said might be displeasing to them; but the blame must be put on the Arabs. The fact was that Adelard had a much broader outlook than most scholars of his time, an outlook widened still more by what he had assimilated of Arabic culture. The orthodox placed implicit reliance upon authority; Adelard was in favour of scientific investigation and set reason above authority. Lynn Thorndike translates as follows the passage in which Adelard strove to make his position clear to his nephew:

I learned from my Arabian masters under the leading of reason; you, however, captivated by the appearance of authority, follow your halter. Since what else should authority be called than a halter? For just as brutes are led where one wills by a halter, so the authority of past writers leads not a few of you into danger, held and bound as you are by bestial credulity. Consequently some, usurping to themselves the name of authority, have used excessive licence in writing, so that they have not hesitated to teach bestial men falsehood in place of truth. For why shouldn't you fill rolls of parchment and write on both sides, when in this age you generally have auditors who demand no rational judgement but trust simply in the mention of an old title? ... Wherefore, if you want to hear anything more from me, give and take reason. For I'm not the sort of man that can be fed on the picture of a beefsteak.

Adelard's main interest was in mathematics, and so far as is known he translated only one Arabic work on alchemy: his importance to our theme is as showing how the spirit of free inquiry was leading men to explore new regions of learning and experience, a movement that gathered strength in the second half of the twelfth century and became irresistible in the succeeding one. Its ambit included the study of alchemy, and the number of Arabic alchemical works translated, often with commentaries, by the year 1300 is in itself so large as to bear striking witness to the success of such scholars as Adelard and Robert of Chester.

The greatest of the translators was Gerard of Cremona, who was born (about 1114) and died (1187) at Cremona, in Lombardy, though some authorities say that he died at Toledo, where he spent most of his life working in the great College of Translators founded by Archbishop Raymond (1126–51). Toledo had been recaptured from the Moors by Alfonso VI in 1105 and became the capital of Castile; but the population still numbered very many Muslims and the language used was still predominantly Arabic. Raymond established his college for the express purpose of making Arabic knowledge available to the West, and attracted thither many European scholars, who were assisted in their work by Arabic-speaking interpreters. The latter rendered Arabic works into Castilian versions, which were then translated into Latin by the Western scholars; though not seldom such of these scholars who had travelled in the East would make Latin translations direct from the Arabic.

No doubt Gerard of Cremona, who does not seem to have gone farther afield than Toledo, at first availed himself of the aid of interpreters, but he was an extremely able linguist and must soon have been able to dispense with intermediaries. Biographers relate that even from infancy Gerard was fascinated by philosophy and learned all that it was possible to learn in every department of it. Coming across references to, or quotations from, Ptolemy's *Almagest* fired him with an ardent desire to read the whole book, but no Latin version of it existed at that time. Gerard therefore determined to go to Toledo and learn Arabic so that he could first study it in that language and afterwards translate it into Latin. In the event, he remained at Toledo for the rest of his life – unless he returned to Cremona to die – engaged in such a spate of translations as perhaps only Hunain ibn Ishaq (p. 67) could have paralleled.

Beholding the abundance of books in every field in Arabic [translates Lynn Thorndike] and the poverty of the Latins in this respect, he devoted his life to the labour of translation, scorning the desires of the flesh, although he was rich in wordly goods, and adhering to the things of the spirit alone. He toiled for the advantage of all both present and future, not unmindful of the injunction of Ptolemy to work good increasingly as you near your end. Now, that his name may not be hidden in silence and darkness, and that no alien name may be inscribed by presumptuous thievery in his translations, the more so since he (like Galen) never signed his own name to any of them, they have drawn up a list of all the works translated by him whether in dialectic or geometry, in astrology or philosophy, in medicine or in the other sciences.

Gerard is credited with having translated a total of no fewer than seventy-six works, some of them, such as Avicenna's 'Canon of Medicine', very lengthy. Of alchemical works, or works indirectly connected with alchemy, he translated a book of Razi's on alums and salts, a book on the properties of minerals, and the first three books of Aristotle's *Meteorologica*, together with the *Almagest* of Ptolemy that took him to Toledo in the first place. It is also believed that he translated a work of Jabir ibn Hayyan, namely the 'Book of Seventy' (p. 74).

It would be easy to expand this list of translators very considerably, but probably enough has been said to explain why all the principal, and many of the minor, alchemical writers and books became familiar so quickly to the West in the twelfth and thirteenth centuries, and why this period has earned the title of the 'Scientific Renaissance'. But in addition to those who undertook the actual work of translation there were other scholars who absorbed Arab learning and showed its influence in their own writings. Such a one was Daniel of Morley – probably Morley in Norfolk – who went to Paris to study Roman law, but found the professors there ponderously ignorant and stupid: 'when they tried to say something I found them most childish'; he therefore travelled on to Spain to sit at the feet of 'wiser philosophers of the universe', finally returning to England 'with an abundant supply of precious volumes' which he used as the basis of his own *Philosophia.* There were also another Englishman, Roger of Hereford, and a learned Jew, Maimonides, who deserve mention here; but alchemy was by now beginning to find European adepts who wrote original books and did not merely adapt Arabic ones, so we may perhaps take leave of the translators at this point.

Before saluting the newcomers, however, we may glance at an aspect of the work of translation that had a lasting effect upon the vocabulary of English. Alchemy, as Robert of Chester observed, was a new science for the West, and one of the difficulties of the translators was that there were no Latin equivalents of many of its technical terms. They chose the easiest, and perhaps the only, way of surmounting this obstacle, namely by a mere transliteration of the Arabic word into Roman letters. Many of these Latinized words then found their way into English versions of the treatises in which they occurred, and in the end became naturalized. The transliterations were not always accurately made – that again would have been impossible, for the Arabic alphabet has more letters than the Latin or English – so that it is sometimes hard to recognize the original form in the final one. The practice of transliteration was not confined to alchemical translations, but was adopted wherever the translator could find no Latin word to correspond; the fact that it was so often resorted to in alchemy demonstrates the unfamiliarity of the subject. Some of the Arabic words thus borrowed in medieval alchemical Latin treatises are as follows; closer transliterations are added in parentheses, and the English equivalents are given. It will be observed that many of the words have now become anglicized:

Abicum (*anbiq*), alembic
Abric (*al-kibrit*), sulphur
Alcalai (*al-qali*), alkali
Alchemy (*al-kimia*), alchemy

Alcazdir (*al-qasdir*), tin; cf. cassiterite
Alchitram (*al-qitran*), pitch
Alcohol (*al-kuhl*), kohl or black eye-paint; but see p. 172.
Almagest (*al-majisti*), almagest
Almizadir (*al-nushadhur*), sal ammoniac
Anticar (*al-tinkar*), tincal, borax
Athanor (*al-tannur*), furnace
Azarnet (*al-zarnikh*), arsenic [sulphides]
Azoth (*al-zauq*), mercury
Carboy (*qarabah*), carboy
Elixir (*al-iksir*), elixir
Heautarit (*utarid*), mercury
Jargon (*jargun*), jargon, a kind of zircon
Luban (*luban*), gum, resin; *luban jawai*, or Javanese resin, was corrupted into benzoin, whence our word benzene
Mattress (*matrah*), heap, cushion
Naphtha (*naft*), naphtha
Natron (*natrun*), natron, whence our symbol Na for sodium
Noas (*nuhas*), copper
Ocob (*uqab*), eagle, sal ammoniac
Tutty (*tutiya*), tutty, zinc oxide
Zaibar (*zaibaq*), mercury
Ziniar (*zinjar*), verdigris.

ALBERTUS MAGNUS AND ROGER BACON

As might be expected, the first results of the abundant influx of knowledge from Islam were to be seen in the sorting out of the new material and in its arrangement for general use. This was done by several compilers, of whom Bartholomew the Englishman and Vincent of Beauvais are the best known. The first of these was a Franciscan who studied at Oxford in the first half of the thirteenth century. He gave a course of lectures on the Bible in Paris, and in 1230 the General of his Order requested that he should be sent to Magdeburg to lecture there also. Bartholomew's great book was entitled 'On the Properties of Things', and was intended for the instruction of the laity and the less well educated among the clergy. It included much earlier learning as well as much of the new, and, as is inevitable in an encyclopaedic work, even in these days, it was somewhat out of date by the time it was completed. Nevertheless it attained great popularity and its success was by no means confined to England. Charles V of France, in 1372, ordered it to be translated into French, and Spanish and Dutch versions quickly followed. It was originally written in Latin, and copies of it in this language could be borrowed by students in Paris on payment of a small fee. It appeared in English in 1397, and as many as seventeen editions of the various versions were published in the course of the fifteenth century.

In the 'chemical' sections of the book, Bartholomew relies largely on Avicenna, and though he has much to say on astrology there is little on alchemy: no doubt he had not made up his mind as to its importance. We may quote what he says about mercury and glass:

Quicksilver is a watery substance medlied strongly with subtle earthly things, and may not be dissolved and that is for great dryness of earth that melteth not on a plain thing. Therefore it cleaveth not to the thing that it toucheth, as doth the thing that is watery. The substance thereof is white: and that is for clearness of clear water, and for the whiteness of subtle earth that is well digested. Also it hath whiteness of medlying of air with the aforesaid things. Also quicksilver hath the property that it curdeth not by itself kindly without brimstone: but with brimstone, and with substance of lead it is congealed and fastened together. And therefore it is said, that quicksilver and brimstone is the element, that is to wit matter, of which all melting metal is made. Quicksilver is matter of all metal, and therefore in respect of them it is a simple element. Isidore saith it is fleeting, for it runneth and is specially found in silver forges as it were drops of silver molten. And it is oft found in old dirt of sinks, and in slime of pits. And also it is made of minium [cinnabar] done in caverns [retorts] of iron, and a patent or a shell done thereunder; and the vessel that is annointed therewith, shall be beclipped with burning coals, and then the quicksilver shall drop. Without this silver nor gold nor latten [brass] nor copper may be over-gilt. And it is of so great virtue and strength, that though thou do a stone of an hundred pound weight upon quicksilver of the weight of two pounds, the quicksilver anon withstandeth the weight. And if thou doest thereon a scruple of gold, it ravisheth unto itself the lightness thereof. And so it appeareth it is not weight, but nature to which it obeyeth. It is best kept in glass vessels, for it pierceth, boreth, and fretteth other matters.

On glass:

Glass, as Avicen saith, is among stones as a fool among men, for it taketh all manner of colour and painting. Glass was first found beside Ptolomeida in the cliff beside the river that is called Vellus, that springeth out of the foot of Mount Carmel, at which shipmen arrived. For upon the gravel of that river shipmen made fire of clods medlied with bright gravel, and thereof ran streams of new liquor, that was the beginning of glass. It is so pliant that it taketh anon divers and contrary shapes by blast of the glazier, and is sometimes beaten, and sometimes graven as silver. And no matter is more apt to make mirrors than is glass, or to receive painting; and if it be broken it may not be amended without melting again. But long time past, there was one that made glass pliant, which might be amended and wrought with an hammer, and brought a vial made of such glass before Tiberius the Emperor, and threw it down on the ground, and it was not broken but bent and folded. And he made it right and amended it with a hammer. Then the Emperor commanded to smite off his head anon, lest that his craft were known. For then gold should be no better than fen [clay], and all other metal should be of little worth, for certain if glass vessels were not brittle, they should be accounted of more value than vessels of gold.

It is clear that Bartholomew's critical faculty is not very acute, and that he is content to accept as true any information for which he can find authority. The section on quicksilver plainly betrays its Arabic origin, and goes back ultimately to Jabir by way of Avicenna. There is as yet nothing new – no original contribution from the West, no fresh theory from European philosophers.

With Vincent of Beauvais, only a few years later than Bartholomew, things are much the same, but more attention is paid to alchemy. Vincent (*c.* 1190–*c.* 1264), a Dominican, was for a time sub-prior of the monastery at Beauvais, near Amiens, where his administrative duties interfered with his literary activities – a fact about which he complains. He was also librarian and chaplain to Louis IX and tutor to the king's two sons, but the numerous demands upon his time still gave him leisure to write a lengthy work entitled *Speculum maius* or 'Great Mirror'. This consists of three parts, the 'Mirror of Doctrine', the 'Mirror of History', and the 'Mirror of Nature', the whole work forming an encyclopedia intended to reflect the sum of contemporary knowledge. The 'Mirror of Nature', says Sarton, 'is an account of natural history in the form of a gigantic commentary on Genesis I'. It is on much the same lines as Bartholomew's encyclopedia, and still fails to give due weight to the new knowledge; but one may observe that the balance is beginning to swing more to that side. Thus Vincent quotes not merely from Avicenna but from other Arabic authors including Razi, Averroës (Ibn Rushd), and Al-Bitruji. His references to alchemical matters are taken mainly from Razi's book on alums and salts (discussed chiefly in the 'Mirror of Doctrine') and from Avicenna. He believed in the main theory of alchemy, namely the possibility of transmutation, and thought that alchemy bore the same relation to mineralogy that agriculture bore to botany. He accepted the sulphur-mercury theory of metals, and thought that base metals could be disintegrated and then rebuilt into gold or silver. His four 'spirits' were mercury, sulphur, arsenic, and sal ammoniac, and he believed that metals are generated and grow in the bowels of the earth.

Such men as Vincent and Bartholomew were, then, not original thinkers but busy selectors and compilers of earlier knowledge; they performed invaluable service but were, so to speak, marking time rather than advancing. In Albertus Magnus we meet a different kind of mind, 'the dominant figure in Latin learning and natural science of the thirteenth century, with whose course his life-time was nearly coincident, the most prolific of its writers, the most influential of its teachers, the dean of its scholars, the one learned man of the twelfth and thirteenth centuries to be called "the Great"'. Albertus Magnus, Count of Bollstädt, was born at Lauingen in Suabia, probably in the year 1193, though some say 1206. Joining the Dominican Order at Padua in 1223, he rapidly became a prodigy of learning and was popularly known as the *Doctor Universalis* – though, like all great men, he had his detractors, who contemptuously nicknamed him the 'Ape of Aristotle'. From 1228 to 1245 he taught at Freiburg, Ratisbon, Strasbourg, and Cologne, while from 1245 to 1248 he lectured in Paris and began the compilation of his great philosophical treatises. He was Provincial of his Order from 1254 to 1257, and bishop of Ratisbon from 1260 to 1262; he then left his episcopal palace on the Danube and retired to a cloister at Cologne to devote the rest of his life to writing and study: a contemporary tells us that Albert found the life of a German bishop too warlike for his taste. He is said to have died at Cologne on 15 November 1280, his intellectual powers having begun to wane some three years earlier. A genuinely pious man, he conformed strictly to the rules of his Order, even walking barefoot on his official journeys through the parts of Germany under his supervision. His fame as a teacher was so great that the young Thomas Aquinas made the long journey from Italy to Cologne to become his pupil.

Albertus – whom tradition describes as a man of exceedingly small stature – was probably the best-read man of his time, and although his reputation rests mainly upon his philosophical works, he had an extensive and unusually accurate knowledge of contemporary science, which he describes in his 'Book on Minerals' and elsewhere. He visited mines, mineral outcrops, and alchemical laboratories, and was a keen field-botanist. Although he yielded to none in his admiration of Aristotle, he would not agree that the great philosopher was either infallible or omniscient, and held that the development of science was not closed by his death. Albertus felt a strong 'desire for concrete, specific, detailed, accurate, knowledge concerning everything in nature', and maintained that, in the study of natural phenomena, a man should not merely transcribe an ancient statement but observe with his own eyes and mind. Yet he does not appear to have appreciated the extreme scientific importance of experiment, as distinct from observation, and though he tested the genuineness of alchemical gold, and offered scraps of iron to ostriches to ascertain whether the old story was true, these were exceptional cases and find few parallels in his writings. Like all his contemporaries he believed in astrology and magic, and, in spite of his own canon of criticism, is often quite ready to admit the fabulous.

Thus he quotes, without expressing disbelief, a story to the effect that if a crow's eggs are boiled and then put back in the nest, the bird will fly off to the Red Sea and return with a stone on contact with which the eggs become raw again. If a man puts this stone into his mouth he can understand the language of birds. Again, Albert reports twins one of whom had such occult power in his right side that all bolts and locks flew open when he turned it towards them; the left side of the other twin, on the contrary, caused all open doors exposed to it to shut. The following experience Albert says he himself witnessed (Lynn Thorndike's translation):

An emerald was recently seen among us, small in size but marvellous in beauty. When its virtue was to be tested, someone stepped forth and said that, if a circle was made about a toad with the emerald and then the stone was set before the toad's eyes, one of two things would happen. Either the stone, if of weak virtue, would be broken by the gaze of the toad; or the toad would burst, if the stone was possessed of full natural vigour. Without delay things were arranged as he bade; and after a short lapse of time, during which the toad kept its eye unswervingly upon the gem, the latter began to crack like a nut and a portion of it flew from the ring. Then the toad, which had stood immovable hitherto, withdrew as if it had been freed from the influence of the gem.

Such tales are common enough in medieval times, and the examples just given are adduced to illustrate the point that even so intelligent a man as Albertus Magnus was ready to accept the marvellous; though in many cases he expresses scepticism or protects himself by saying merely that such and such a tale was reported to him. This belief in a magical background is inherent in most if not all medieval alchemy and helps to explain the often bizarre procedures followed by the adepts.

Albertus was not an Arabic scholar, but was well acquainted with Latin translations of Avicenna, Averroës, and other Muslim writers. In his 'Book of Minerals' he moulds his views on alchemy very largely in accordance with Avicenna's opinions expressed in the chapters from 'The Remedy' translated about 1200 by Alfred the Englishman (p. 94). Thus he believes that most alchemists merely succeeded in dyeing metals so that they resemble gold or silver, the actual metallic species remaining unaltered. 'Alchemy,' he says, 'cannot change species but only imitate them.... I have myself tested alchemical gold and found that after six or seven ignitions it was converted into powder.'

Perhaps, however, in this passage he is referring only to the generality of the alchemists, for in another book, 'The Little Book of Alchemy', he relates that he was given a knowledge of alchemy by the grace of God. It is true that the authenticity of this book is not definitely established, but it was ascribed to Albertus before 1350. The author recounts the errors of his predecessors, and promises to describe nothing but what he has actually seen. Next he states eight rules to be observed by the alchemist, much in the style of an admonition made five centuries earlier by Jabir. He then proceeds to discuss the various operations and pieces of apparatus used in alchemy, and describes the common chemical substances and experiments that may be carried out with them. Finally, recipes are given for the production of gold and silver. The belief is expressed 'that metals can be produced by alchemy which are the equal of natural metals in almost all their qualities and effects', except that alchemical iron does not posses magnetic properties and that alchemical gold lacks certain curative powers supposed to inhere in the natural metal.

Albert's almost equally famous pupil, Thomas Aquinas, believed like his master in the possibility of making gold and silver alchemically, though he says that the art is a difficult one. He returns to Aristotle in holding that metals arise ultimately from a dry or smoky exhalation of the earth, and a moist or watery one; though these exhalations at first change into sulphur and mercury respectively. But Thomas thinks that the generation of metals requires also the occult operations of a celestial virtue not always under alchemical control, so that the aim of the worker should be to arrange conditions under which the virtue will be likely to function.

With Roger Bacon we come to another medieval scholar of very wide learning, and one who captured the popular imagination enough to set going a large number of highly picturesque legends: he could 'make women of devils and juggle cats into costermongers'. This vulgar conception of Bacon as a sorcerer and necromancer was satirized by Robert Greene (1560–92) in his play 'The Honorable Historie of Frier Bacon and Frier Bongay', performed in 1594. Here Bacon and his companion wizard Bongay construct a brazen head, and with the Devil's help propose to endow it with speech. When all is prepared, the head is due to give utterance within a month, but their labour would be lost 'if they heard it not before it had done speaking'. Bacon watched day and night for three weeks but was then so overcome by sleepiness that he handed over to his servant Miles. Only a little while afterwards the head said 'Time is', but Miles feared to wake his master for so little. The head then said 'Time was', relapsed into silence, and finally ejaculated 'Time is past'; with that it broke to pieces, the noise awakening Bacon, who roundly cursed his servant. Such stories are like snowballs in gathering bulk as they roll along, and those about Bacon are no exception. Even within the last few years it was reported that 'an ancient manuscript' in cipher had been discovered 'in a castle in southern Europe', from which it appeared that Bacon had invented the compound telescope and microscope and had observed the nuclei of cells. The Middle Ages did not have a monopoly in credulity.

Roger Bacon was born at Ilchester in Somerset, probably in 1214, and 'appears to have belonged to a wealthy family, which, subsequently, in the struggle between Henry III and the Barons (1258–65), sacrificed their fortunes in the cause of the King'. Under the influence of the English philosopher and scholar Robert Grosseteste, bishop of Lincoln, he undertook the study of Greek at Oxford, and it was doubtless Grosseteste who persuaded him to join the Franciscan Order about 1247. From 1234 to 1250 he studied and lectured at the university of Paris, choosing as his master 'one of the most modest and most learned men of the time, one who had devoted himself to the study of chemistry and mathematics and astronomy and, above all, to those practical applications of experimental science which prompted his enthusiastic pupil to call him "the Master of Experiments"' – namely Petrus Peregrinus, author of one of the first treatises on the magnet.

Between 1250 and 1257 Bacon probably spent most of his time at Oxford, but in the latter year, having fallen under the suspicions of the authorities, or perhaps, as he says himself, because of poor health, he took no part in the outward affairs of the university and remained in such retirement for another decade. Later his criticism of authority, independence of thought, and general quarrelsomeness again brought him into conflict with his superiors in the Order, and it is supposed that he was kept in confinement in Paris for fourteen years (1277–91). In 1292 he was set at liberty and returned to Oxford; but his freedom was short-lived, for 'the noble doctor Roger Bacon was buried at the Grey Friars church of the Franciscans [long demolished] in Oxford, A.D. 1292, on the Feast of St Barnabas the Apostle [11 June]'. A tower, traditionally known as 'Friar Bacon's Study', stood until 1779 on Folly Bridge, on the south side of Oxford.

In 1267 Bacon writes, in his *Opus Tertium*:

I have laboured from my youth in the sciences and languages, and for the furtherance of study, getting together much that is useful. I sought the friendship of all wise men among the Latins, and caused youth to be instructed in languages and geometric figures, in numbers and tables and instruments, and many needful matters. I examined everything useful to the purpose, and I know how to proceed, and with what means, and what are the impediments: but I cannot go on for lack of the necessary funds. Through the twenty years in which I laboured specially in the study of wisdom, careless of the crowd's opinion, I spent more than two thousand livres [about £10,000] in these pursuits on occult books (*libros secretos*) and various experiments, and languages and instruments, and tables and other things.

Bacon was indeed a 'devotee of tangible knowledge', but he was not so far in advance of his time as has sometimes been said. In common with all other Christians of his age, he believed that the Bible contained, either explicitly or implicitly, the whole realm of knowledge. On the other hand, to understand the Bible thoroughly every art and science is necessary – though the patriarchs and prophets had full knowledge of all sciences, magic and astrology included. The queen of sciences is, therefore, theology, and all other branches of learning are her handmaids. Round this central theme Bacon's whole system continually revolves, and we cannot understand his attitude to natural philosophy if we forget this cardinal fact. His advocacy of the experimental method was therefore primarily concerned not with the search for objective truth but with the exposition of scriptural scientific knowledge, and it is only within these limits that it must be envisaged. Bacon, in short, must be judged against the intellectual background of his day, and must not be gratuitously credited with a mental outlook that in actual fact arose only very much later. Moreover, by 'experience' Bacon implied something more than observation and experiment; for him it included the illumination of faith, spiritual intuition, and divine inspiration, and this esoteric experience was 'much better' than the experience of philosophy or science.

Bacon distinguished two kinds of alchemy, namely 'speculative' and 'practical'. Practical alchemy he regarded as more important than the other sciences, as more productive of material advantages than they. Speculative alchemy

treats of the generation of things from the elements and of all inanimate things and of simple and composite humours, of common stones, gems, marbles, of gold and other metals, of sulphurs and salts and pigments, of lapis lazuli and minium and other colours, of oils and burning bitumens and other things without limit, concerning which we have nothing in the books of Aristotle. Nor do the natural philosophers know of these, nor the whole assembly of Latin writers. And because this science is not known to the generality of students it necessarily follows that they are ignorant of all that depends upon it concerning natural things, namely of the generation of animate things, of plants and animals and men, for being ignorant of what comes before they are necessarily ignorant of what follows.

But there is another alchemy, operative and practical,

which teaches how to make the noble metals, and colours, and many other things better or more abundantly by art than they are made in nature. And the science of this kind is greater than all those preceding because it produces greater utilities. For not only can it yield wealth and very many other things for the public good, but it also teaches how to discover such things as are capable of prolonging human life for much longer periods than can be accomplished by nature.... It confirms theoretical alchemy through its works and therefore confirms natural philosophy and medicine, and this is plain from the books of the physicians. For these authors teach how to sublime, distil and resolve their medicines, and by many other methods according to the operation of that science, as is clear in health-giving waters, oils and many other things.

These passages throw light on Bacon's conception of the role of experiment, using the word in our modern sense. Experiments are not used as a basis for inferring general laws from particular cases, as with us; on the contrary they are to serve as confirmation of conclusions reached by deduction from general principles already accepted. This is so much the usual habit of alchemy as to be worth bearing constantly in mind when attempting to understand alchemical reasoning.

Except for the fact that he minimized the importance of the Aristotelian 'prime matter' and made fuller use of the theory of the four elements, Bacon differs very little from the other alchemists in his ideas about metallic constitution and transmutation. He accepts the sulphur-mercury theory, which he appears to have taken directly from Avicenna, and is quite as credulous about the manufacture of gold as any of his contemporaries. He had a wide acquaintance with the Arabic authors, whom, unlike Albertus Magnus, he could read in the original, and he seems to have felt that in alchemy there might be a link between Aristotelian physics and the biological sciences. There is no evidence that he made any discoveries in practical alchemy, and the affirmation that he invented gunpowder is completely unfounded; he did, however, suggest that if gunpowder were enclosed in a solid container its explosive effect would be greater.

Bacon also foresaw a number of ways in which science might exert mastery over nature, and here is his vision of the future:

Machines for navigation can be made without rowers so that the largest ships on rivers or seas will be moved by a single man in charge with greater velocity than if they were full of men. Also cars can be made so that without animals they will move with unbelievable rapidity; such we opine were the scythe-bearing chariots with which the men of old fought. Also flying machines can be constructed so that a man sits in the midst of the machine revolving some engine by which artificial wings are made to beat the air like a flying bird. Also a machine small in size for raising or lowering enormous weights, than which nothing is more useful in emergencies. For by a machine three fingers high and wide and of less a man could free himself and his friends from all danger of prison, and rise and descend. Also a machine can easily be made by which one man can draw a thousand to himself by violence against their wills, and attract other things in like manner. Also machines can be made for walking in the sea and rivers, even to the bottom without danger. For Alexander the Great employed such, that he might see the secrets of the deep, as Ethicus the astronomer tells. These machines were made in antiquity and they have certainly been made in our times except possibly a flying machine which I have not seen nor do I know of any one who has, but I know an expert who has thought out the way to make one. And such things can be made almost without limit, for instance, bridges across rivers without piers or other supports, and mechanisms, and unheard-of engines.

Bacon also suggested that imperfect sight could be aided by the use of suitably shaped lenses, urged a reform of the calendar, and foresaw the possibility of circumnavigating the globe. As Sarton remarks, 'One such anticipation would hardly deserve to be mentioned, but the combination of so many in a single head is very impressive.'

ARNOLD OF VILLANOVA

Few names carried more weight in early European alchemical circles than that of Arnold of Villanova, a Catalan, who was born near Valencia about 1235. He seems to have come of a poor family, for he speaks of his youth as a time of hardship, but he took full advantage of the education given him by Dominicans and then went on to study medicine at Naples. He also learnt Arabic and some Greek as well as Latin, which was the common tongue, at least in writing, of all educated Europeans at that time. His genius at medicine soon brought him fame, and he was frequently summoned to treat popes and kings and other notabilities; as a result he travelled extensively in Spain, France, Italy, and even North Africa. This, however, did not prevent him from writing a great many books, in such various places as Barcelona, Bologna, Naples, Rome, and Valencia, and his literary output augmented his fame. In 1285 he was called to attend the dying King Peter III of Aragon, and was rewarded with the gift of a castle in Tarragona and a professional chair in the University of Montpellier.

But Arnold was more than a physician: he was also an

astrologer, an alchemist, a controversialist, a social reformer, and a diplomat, while at Montpellier it was apparently not long before he became the chief power in the university. His attacks on abuses among the clergy, and his frank expression of original ideas on theological subjects, made him obnoxious to some of the authorities of the Church, and when he was sent on a diplomatic mission to Paris in 1299, by James II of Aragon, he was arrested by minions of the Holy Office. He was released on bail the next day, but had to stand trial in Paris for having written a book, considered heretical, on the impending arrival of the Antichrist. In spite of vigorous protests to the King of France (Philip the Fair) and to Pope Boniface VIII, he was not allowed to leave Paris until 1301, when he went to Genoa.

His troubles were not over, however. He had sent a carefully modified version of his book on Antichrist to Boniface, who could not find much in it to reprehend; but his enemies in Paris countered by letting the Pope have a copy of the original, whereupon Arnold was imprisoned again. Fortunately for him, Boniface about this time developed an attack of the stone, of which Arnold was able to cure him, so things turned out happily after all. Boniface released him and even went so far as to present him with a castle at Anagni – where, by the irony of fate, the Pope was himself imprisoned a little later by agents of Philip the Fair, and where he died.

After his release, and presumably after Boniface's death in 1303, Arnold went to Marseilles and then on to Barcelona, where in 1305 he made a will giving instructions for the disposal of his library, for charitable bequests, and for provision for his wife and children: but the precaution was premature and he still had six years to live. These were spent in performing numerous services for reigning monarchs and in writing further books on medicine and alchemy. He continued to do a good deal of travelling, for during this time he visited, among other places, Rome, Avignon, Montpellier, Bordeaux, and Naples, where he met Ramon Lully (p. 126). Returning by ship from Naples to Genoa, he died at sea in 1311.

Arnold of Villanova presents us with a curiously mixed philosophy. In the first place he appreciated the importance of the study of natural science and urged that it should be given a greater consideration in education. He also emphasized the value of experiment and bemoaned the difficulty of making experiments in medicine. On the other hand, although he sometimes discountenanced magic his works abound in superstitious ideas, and he saw nothing illogical in trying to cure disease by magic gems and by the influence of the stars. Even his treatment of Boniface VIII for the stone included the application of a seal in the form of a lion, and he wrote a treatise on seals and amulets. After giving details of how one such seal should be engraved, he goes on to enumerate its virtues as follows (Lynn Thorndike's translation): 'This precious seal works against all demons and capital enemies and against witchcraft, and is efficacious in winning gain and favour, and aids in all dangers and financial difficulties, and against thunderbolts and storms and inundations, and against the force of the winds and the pestilences of the air. Its bearer is honoured and feared in all his affairs. No harm can befall the building or occupants of the house where it is. It benefits demoniacs, those suffering from inflammation of the brain, maniacs, quinsy, sore throat, and all diseases of the head and eyes, and those in which rheum descends from the head. And in general I say that it wards off all evils and confers good; and let its bearer abstain as far as possible from impurity and luxury and other mortal sins, and let him wear it on his head with reverence and honour.' If this portable all-in policy fails in cases of mania, however, there is nothing for it but to pierce a hole in the skull in order to allow the noxious vapours affecting the brain to escape; while as for failing sight, this may be due to nothing more than too frequent washing of the head.

Such astonishing credulity leads us not to expect any great lucidity in Arnold's treatment of alchemy. He wrote several works on the subject, and many more have been wrongly attributed to him. The longest of those that are probably genuine is called 'The Treasure of Treasures, Rosary of the Philosophers, and Greatest Secret of all Secrets'; it consists of two parts dealing with theory and practice respectively. It became very popular and set the fashion for very many later alchemical 'Rosaries'.

Like many alchemists, Arnold sets out by saying that he will conceal nothing and keep nothing back, but that the reader should be prepared for hidden reasoning and should also supplement the 'Rosary' with a perusal of, and reflection upon, other books; but Arnold believes that he himself has divined the secret common to Plato, Aristotle, and Pythagoras. He accepts the sulphur-mercury theory of metallic constitution, but regards mercury as much the more significant of the two; in fact, ordinary sulphur is harmful to metals, and the sulphur that is envisaged by Arnold is that already hidden in the mercury. Hence it should be possible to prepare gold and silver from mercury alone, though with the introduction of a little of the precious metal required in order to start the reaction and, as we should say, catalyse it. Lead cannot be substituted for mercury, and the best mercury or 'mercurial liquid' is that exported from Spain in containers sealed with the Spanish seal.

It is not clear what this 'mercurial' *aqua vitae* is, but Arnold says that for transmuting base metals it should be used in the proportion of four of liquid to one of metal, whereas for preparing the elixirs the proportion should be twelve of liquid to one of gold or silver. The elixirs so obtained should be capable of transmuting one thousand times their own weight of base metal into the precious ones, fit to withstand full assay. Other methods of obtaining gold from mercury are described, including one that involves the separation of the mercury into the four elements by means of a ferment, and then recombining them in the shape of gold. For this process to succeed, the elements to be recombined must be in definite proportions by weight: thus gold requires the ratio of 1 : 1 water and air, 2 : 1 water and fire, and 3 : 2 water and earth; it is also necessary to know the relative heat, cold, dryness, and moistness. As usual, in all these processes colour-changes must be carefully watched, the formation of a red impalpable powder being the sign of completion and success in the preparation of the great elixir. Arnold adds a warning that when projecting the elixir it must not be dissolved or fused, possibly because if heated it would vaporize and so be lost.

It would be difficult to equate these statements with any

known chemical facts, but, although Arnold compares the alchemical work with the conception, birth, crucifixion, and resurrection of Christ, there must have been a practical experimentation somewhere beneath the complex, involved, and often self-contradictory mass of theory overlying all. Alchemists were inclined to extrapolate from what they had actually observed to what they thought might or ought to have happened if their money had not run out or if their glass 'egg' had not broken at the critical moment. They were unexcelled in bearing out Pope's aphorism:

> Hope springs eternal in the human breast:
> Man never is, but always to be, blest.

That Arnold did in fact carry out some practical chemistry is proved by his account of the dry distillation of human blood from an alembic. He says that the first fraction of the distillate is a clear water, and this is the element water. Soon, however, a yellowish liquid begins to come over, and at this point the receiver should be changed. The yellow distillate is the second element, namely air. After the yellow oil comes a red one, for which a third receiver should be employed; this third distillate is the element fire. Now this description does correspond with the facts, though not with their interpretation, and is of interest as an early account of fractional distillation. But Arnold cannot refrain from adding that the 'fire' obtained from blood in this way has remarkable medicinal virtues; thus a certain count lay at the point of death, but administration of the 'fire' revived him sufficiently to make his last confession before he died. To his credit, Arnold does not suggest that any portion of the distillate can effect transmutation – and further to his credit is that he seems to have been the first to observe the poisonous character of carbon monoxide.

RAMON LULLY

In medieval alchemical literature the name of Arnold of Villanova is often coupled with that of Ramon Lully (about 1232–1315), another Catalan philosopher and a Christian missionary to the Moors. He was for a time tutor to the sons of James I (*El Conquistador*, King of Aragon from 1213 to 1276), and afterwards spent nine years in Majorca studying Arabic with a Moorish slave. He made three missionary journeys to North Africa, on the last of which he was stoned to death. Though a number of books on alchemy are ascribed to him, none of them is authentic, and it is well known that he disbelieved in the possibility of transmuting one metal into another. Nevertheless the glamour of his name was sufficient for the alchemists to claim him as one of their company, and he was alleged not merely to be able to transform himself into a red cock at will but to have turned twenty-two tons of base metal into gold in the Tower of London in order to enable King Edward III (or II) of England to equip a crusade against the Turks. This story, which is entirely fabulous, is linked with the name of an equally fabulous John Cremer, a supposed Abbot of Westminster. In his *Testament*, 'Cremer' describes himself as a very earnest follower of the Art, but greatly holden back by the obscurity of the books he read on it; he studied them for thirty years, and at much expense tried to follow their precepts with practice, but he lost both time and labour. The more he read, the more he erred, until it pleased God to bring him to Italy, where he met Ramon, in whose company and fellowship he stayed long, to the end that he should open some part of the great mystery. Furthermore, by persistent entreaty he got Ramon to return with him to England, where he stayed in Cremer's house for two years and admitted Cremer to the secret.

Cremer then claims to have brought this excellent man unto the sight of King Edward, by whom he was most worthily received and kindly entertained, and allowed himself to be persuaded to make a supply of gold by alchemical means. He imposed the conditions that the king in his own person should fight against the Turk, that some of the gold should be bestowed upon the house of Our Lord, and that the king should not expend any of it in pride or in warring against fellow-Christians. But the king broke his promise, and the holy man was sore afflicted in spirit, apparently reproaching the faithless one so vehemently that he found himself 'clapt up in the Tower'. Here, according to Ashmole (p. 189), he made himself a 'leaper' – presumably a leaping-pole – by means of which he escaped, afterwards crossing safely to France.

From the alchemical gold, Edward was said to have had rose nobles struck – though such coins were actually not minted before 1465 – and even as late as 1696 a diarist notes having seen 'a rose noble, one of those that Ramund Lully is sayd to have made by chymistry'. The reverse of the coins was inscribed with the words: '*Iesus autem transiens per medium eorum ibat*, that is, as Jesus passed invisible and in most secret manner by the midst of the *Pharises*, so that *Gold* was made by *invisible* and *secret Art amidst* the *Ignorant*'. If we may believe Ashmole, Cremer had an alchemical device painted on a wall in Westminster Abbey, though it was later washed over with plaster; it depicted in graphic form 'the Grand *Misteries* of the *Philosophers Stone*'.

Lully's rose nobles were by no means the only contributions of alchemy to numismatics; many other alchemical coins and medals have been described by H. C. Bolton. Christian IV of Denmark (1577–1648) employed an alchemist named Kaspar Harbach to transmute base metals into gold, and ducats of 1644 and 1646 were struck from this alchemical product. Some people were sceptical about the origin of the gold, however, so in 1647 the king had further ducats coined to vindicate the honour of his alchemist. These bear on the obverse a full-length figure of the king in armour, with the words: *Christianus D.*[*ei*] *G.*[*ratia*] *Dan.*[*iae*] *R.*[*ex*]. The reverse bears a pair of spectacles and the legend *Vide Mira Domi* [*ni*]. 1647. [See the wonderful works of the Lord.]

In the same year an adept named J. P. Hofmann carried out a transmutation in the presence of the Emperor Ferdinand III, at Nuremberg. From this hermetic gold the emperor caused a medal of rare beauty to be struck. It bears on the obverse two shields, in one of which are eight fleurs-de-lys, and in the other is a crowned lion. The Latin inscriptions signify, 'The yellow lilies lie down with the snow-white lion; thus the lion will be tamed, thus the yellow lilies will flourish', and that the metal was made by Hofmann. A further inscription reads *Tincturae*

Guttae V Libram, denoting that five drops of the tincture or elixir transmuted a whole pound of the base metal. On the reverse is a central circle with Mars in it, holding the symbol ♂ in one hand and a sword in the other. Around this central circle are six smaller ones, containing the signs of gold, silver, copper, lead, tin, and mercury, with an inscription claiming that in this case the active agent in transmutation was made from iron.

Not long afterwards, in January 1648, Ferdinand III was dabbling in alchemy again. A certain Richthausen, who claimed to have received the secret of the Art from an adept now dead, performed a transmutation in the presence of the emperor and of the Count von Rutz, director of mines. All precautions against fraud were taken, yet with one grain of the powder provided by Richthausen two and a half pounds of mercury were changed into gold. To commemorate this event the emperor had a medal struck of the value of 300 ducats. The obverse bore a full length representation of Apollo with rays proceeding from his head; his feet were covered with winged sandals to indicate the conversion of mercury into gold. The inscription read (in Latin), 'The Divine Metamorphosis, exhibited at Prague, 15 January 1648, in the presence of his Imperial Majesty Ferdinand III'. The reverse was plain, except for an inscription to the effect that 'Like as rare men have this art, so cometh it very rarely to light. Praised be God for ever, who doth communicate a part of His infinite power to us His most abject creatures.'

In 1650 the emperor himself made a transmutation with some of Richthausen's powder, and again had a medal struck, this time with the inscription *Aurea progenies plumbo prognata parente*, 'A golden daughter born of a leaden parent.' In token of his admiration and gratitude, Ferdinand now raised Richthausen to the nobility. The new Count appears once more in 1658, when he gave the Elector of Mainz some of his elixir, with which a successful transmutation of mercury into gold was made.

Bolton also tells of an Augustinian monk named Wenzel Seyler, a native of Bohemia, who visited Vienna in 1675 and was granted an interview by the Emperor Leopold I, son of Ferdinand III. Seyler accomplished a transmutation in the royal presence, converting a copper vessel into gold and changing tin into gold. From the latter gold ducats were struck, bearing a portrait bust of the emperor on the obverse, and on the reverse the date 1675 and the lines:

Aus Wenzel Seyler's Pulvers Macht
Bin Ich von Zinn zu Gold Gemacht.

That is, 'By the power of Wenzel's powder I was made into gold from tin.'

Like Richthausen, Wenzel Seyler was rewarded by being ennobled under the title of Wenzel von Reinburg, but owing to some deceits that he subsequently practised he was banished for a time to his cloister. Two years later he was restored to favour and performed another transmutation; a medallion struck from the product was still in existence in the Imperial Cabinet of Coins, Vienna, in 1888, and was examined there by Bolton himself. He says it was of elaborate workmanship but decidedly brassy in colour; moreover its specific gravity was only 12·67 whereas that of gold is 19·3.

More alchemical medals derive from the shady operations of Christian William Krohneman, who actually survived as court alchemist to the Margrave George William of Bayreuth for the long period of nearly ten years, 1677–86. He claimed to be able to transmute mercury into gold, and the Margrave was credulous enough to support him in a lavish style of living and to advance large sums of money to defray the cost of experiments. When the experiments had proved unsuccessful after repeated efforts, Krohneman's reputation began to suffer; but then, in the presence of the Margrave, he heated mercury with salt, vinegar, and verdigris in an iron dish, and at the end of the operation gold remained. Silver also was made alchemically, and a medal struck from it has on the obverse a figure of the winged Mercury; the reverse has a Latin inscription translated by Bolton as follows: 'Let no one be ignorant of the fact that what many have believed to be the work of nature alone is not less the work of art. They were formerly produced, they are now produced, as shown by the thing itself. To the glory of God, the salvation of mankind, and the admiration of the whole world.'

Krohneman had thus retrieved his fortunes and was made a Baron. The Margrave also bestowed monetary and other favours upon him, but the Baron was full of guile and swindled many persons in authority, including General Kaspar von Lilien, whom he fleeced to the extent of 10,000 gulden. During the period of his ill-gotten prosperity Krohneman performed other transmutations leading to alchemical medals, but being at length detected in undeniable fraud he was hanged by order of the Margrave.

A somewhat similar adventurer to Krohneman was Domenico Manuel Caetano, who though of Neapolitan peasant origin managed to pass himself off as a Count and ultimately rose to high office in Germany and Austria. He had been apprenticed to a goldsmith and also acquired virtuosity as a conjuror – an ominous combination. According to his own account he was sufficiently fortunate to discover an unknown alchemist's hidden treasure, together with a manuscript describing the preparation of the philosophers' stone. He began to demonstrate transmutations and did this so spectacularly in Madrid that the Bavarian ambassador there invited him to go to Brussels and exhibit his powers to Maximilian Emanuel of Bavaria, who was Governor of the Spanish Netherlands. The Governor believed Caetano's claim to possess the secret, and advanced him altogether 60,000 gulden, upon which Caetano tried to decamp. Unluckily for him, he was captured and imprisoned for six years; but a second attempt at escape was successful and in 1704 Caetano was in Vienna, where he performed certain transmutations before the Emperor Leopold I and gained the confidence of the whole court. From Vienna he went to Berlin, where he conducted successful experiments in the presence of Konrad Dippel – who was not impressed – and promised King Frederick I to make him a large quantity of the philosophers' stone within sixty days. To explain the ease with which these monarchs parted with their money, it must be remembered that a belief in material alchemy was still very widely held; in any case Frederick gave the adept numerous and valuable presents and appointed him to lucrative offices. When the supply of the Stone was not forthcoming at the end of the sixty days, the king grew restive and Caetano fled to Hamburg; but his freedom was short-lived and he was arrested and conveyed to the fortress at Küstrin (Kostrzyn). Protesting that he could not work in prison, he was taken to Berlin, whence he escaped to Frankfurt-am-Main, only to be re-arrested and sent back to Küstrin. It had now become quite clear that he was an imposter, and in August 1709, dressed in a cloak covered with glittering tinsel, he was hanged from a gilded gallows. It seems only poetic justice that a medal was struck to commemorate the occasion.

A few years before Caetano came to his inglorious end, Georg Stolle, a goldsmith of Leipzig, was visited by a stranger who, after a few opening remarks, asked him if he knew how to make gold. Stolle replied cautiously that he knew only how to work with that metal when already made. The stranger then asked him whether he believed in the possibility of transmutation, to which the goldsmith answered that he believed in the Art of Hermes but had never seen a transmutation performed. The mysterious visitor then produced an ingot which the goldsmith tested and found to be 22-carat gold; he was assured by its owner that it was of alchemical origin. Next day the visitor brought the ingot back again and asked that it should be cut into seven round pieces; this Stolle did, and after the stranger had stamped them he gave him two of the pieces as a souvenir. The inscription stamped on them said (in Latin): 'O Thou who art Alpha and Omega, Thou art the hope of life after death. The restoration of life to lead transforms it into gold and silver. O unequalled love of God in Three Persons, have mercy on me through eternity. By sulphur, salt, and mercury the philosophers' stone is made.' The news of this peculiar incident spread rapidly, and the King of Poland took possession of one of the pieces while the other was deposited in the collection of medals at Leipzig. Nothing further was ever heard of the adept.

At about the same time (1705) a remarkable event occurred in Sweden. The Swedish general Paykhull had been convicted of treason and sentenced to death. In an attempt to avert this punishment, he offered the king, Charles XII, a million crowns of gold annually, saying that he could make it alchemically; he claimed to have received the secret from a Polish officer named Lubinski, who had himself obtained it from a Corinthian priest. Charles accepted the offer, and a preliminary test was arranged under the supervision of a British officer, General Hamilton of the Royal Artillery, as an independent observer. All the materials were prepared with great care, in order to prevent the possibility of fraud, then Paykhull added his elixir and a little lead, and a mass of gold resulted which was coined into 147 ducats. A medal struck at the same time bore the inscription (in Latin as usual): 'O. A. von Paykhull cast this gold by chemical art at Stockholm, 1706.'

Another story of this kind comes from France, where an ignorant Provençal rustic named Delisle caused a sensation by claiming, with apparent good reason, that he could transform iron and steel into gold. The news came to the ears of the Bishop of Senez, who after witnessing one of Delisle's experiments wrote to the Minister of State and Comptroller-General of the Treasury in Paris that he could not resist the evidence of his senses. In 1710 Delisle was summoned to Lyons, where, in the presence of the Master of the Lyons Mint, he made much show of distilling some unknown yellow liquid. He then projected two drops of the liquid upon three ounces of pistol bullets fused with saltpetre and alum, and poured the molten mass out on to a piece of iron armour, where it appeared as pure gold, withstanding all tests. The gold thus obtained was coined by the Master of the Mint into medals inscribed *Aurum arte Factum*, 'Gold made by Art', and these were deposited in the museum at Versailles. Of Delisle's subsequent life, history has nothing to relate.

Finally, mention may be made of a medal on the reverse of which is a Latin inscription rendered by Bolton as 'In the month of July, 1675, I, Doctor J. J. Becher, transmuted by hermetic art this ounce of purest silver from lead.' This is of particular interest, inasmuch as Becher was an eminent German chemist and physician still honoured (or execrated) in chemical circles as, jointly with G. E. Stahl, founder of the phlogiston theory of combustion. Although a learned and skilful chemist, Becher was a firm believer in alchemy, and in fact in 1673 made a proposition to the States General of Holland to manufacture for the government one million thalers of gold a year, over and above expenses, by operations on sea-sand. He maintained that when the sand was fused with certain ingredients and silver, it yielded one thaler of gold for every mark of silver employed. The government accepted the offer, and terms were settled by which Becher was to receive a premium and a percentage of the profits. A preliminary small-scale trial was successful, but shortly afterwards Becher left the country and the scheme was never proceeded with.

'Lully's rose nobles' thus began a curious passage in the history of numismatics, a passage in which there were plenty of comedy, unlimited credulity, and not a little tragedy.

THE LATIN GEBER

It has already been mentioned (p. 68) that the name of Jabir ibn Hayyan occurs in Western alchemy as Geber; variants are Ieber (indicating that the G was soft, corresponding in pronunciation with the Arabic J) and Geber ebn Haen (proving the identity). One or two of Jabir's Arabic works exist in medieval Latin translation (p. 74), but other works in Latin that pass under his name are without known Arabic originals. In English, these works are entitled 'The Sum of Perfection', 'The Investigation of Perfection', 'The Invention of Verity', 'The Book of Furnaces', and 'The Testament'. The most important of them is 'The Sum of Perfection', of which the oldest Latin manuscripts date from the late thirteenth century.

The question at once arises whether the Latin works are genuine translations from the Arabic, or whether they were written by a Latin author and, according to a common practice, ascribed to Jabir in order to heighten their authority. That they are based on Muslim alchemical theory and practice is not questioned, but the same may be said of most Latin treatises on alchemy of that period; and from various turns of phrase it seems likely that their author could read Arabic. But the general style of the works is too clear and systematic to find a close parallel in any of the known writings of the Jabirian *corpus*, and we look in vain in them for any references to the characteristically Jabirian ideas of 'balance' (p. 76) and alphabetic numerology (p. 77). Indeed, for their age they have a remarkably matter-of-fact air about them, theory being stated with a minimum of prolixity and much precise practical detail being given. The general impression they convey is that they are the product of an occidental rather than an oriental mind, and a likely guess would be that they were written by a European scholar, possibly in Moorish Spain. Whatever their origin, they became the principal authorities in early Western alchemy and held that position for two or three centuries.

In 'The Investigation of Perfection' Geber gives a concise account of basic alchemical theory, which may be quoted here in English from Richard Russell's translation (1678):

This *Science* treats of the *Imperfect Bodies* of *Minerals*, and teacheth how to perfect them; we therefore in the first place consider *two Things, viz. Imperfection* and *Perfection*. About these two our Intention is occupied, and of them we purpose to treat. We compose this *Book* of *Things perfecting* and *corrupting* (according as we have found by experience) because Contraries set near each other, are the more manifest.

The *Thing* which perfects in *Minerals*, is the substance of *Argentive* and Sulphur proportionably commixt, by long and temperate decoction in the Bowels of clean, inspissate, and fixed *Earth* (with conservation of its *Radical Humidity not corrupting*) and brought to a solid fusible Substance, with due Ignition, and rendered Malleable. By the *Definition* of this *Nature* perfecting, we may more easily come to the Knowledge of the *Thing* corrupting. And this is that which is to be understood in a contrary Sense, *viz.* the pure *substance* of *Sulphur* and *Argentvive*, without due Proportion commixed, or not sufficiently decocted in the Bowels of unclean, not rightly inspissate nor fixed *Earth*, having a Combustible and Corrupting *Humidity*, and being of a rare and porous Substance; or having Fusion without due Ignition, or no Fusion, and not sufficiently Malleable.

The first Definition I find intruded in these two Bodies, *viz.* in *Sol* and *Luna*, according to the Perfection of each; but the second in these four, *viz. Tin, Lead, Copper* and *Iron*, according to an Imperfection of each. And because these Imperfect Bodies are not reducible to *Sanity* and *Perfection*, unless the contrary be operated in them; that is, the Manifest be made Occult, and the Occult be made Manifest: which Operation, or Contrariation, is made by Preparation, therefore they must be prepared, Superfluities in them removed, and what is wanting supplied; and so the known Perfection inserted in them. But Perfect Bodies need not this preparation; yet they need such Preparation, as that, by which their Parts may be more Subtiliated, and they reduced from their *Corporality* to a fixed *Spirituality*. The intention of which is, of them to make a Spiritual fixed Body, that is, much more attenuated and subtiliated than it was before. . . .

We find Modern Artists to describe to us one only *Stone*, both for the *White* and for the *Red*; which we grant to be true: for in every *Elixir*, that is prepared, *White* or *Red*, there is no other Thing than *Argentvive* and *Sulphur*, of which, one cannot act, nor be, without the other: Therefore it is called, by *Philosophers*, one *Stone*, although it is extracted from many Bodies or Things. For it would be a foolish and vain thing to think to extract the same from a Thing, in which it is not, as some infatuated Men have conceited; for it was never the Intention of *Philosophers*: yet they speak many things by similitude. And because all *Metallick* Bodies are compounded of *Argentvive* and *Sulphur*, pure or impure, by accident, and not innate in their first Nature; therefore, by convenient *Preparation*, 'tis possible to take away such Impurity. For the *Expoliation* of *Accidents* is not impossible: therefore, the end of *Preparation* is, to take away *Superfluity* and supply the *Deficiency* in Perfect Bodies. But *Preparation* is diversified according to the *Diversity* of things indigent. For experience hath taught us diverse ways of acting, *viz. Calcination, Sublimation, Descension, Solution, Distillation, Coagulation, Fixation*, and *Inceration*.

Incorrect as are the premisses from which he starts, Geber nevertheless explains them clearly and argues from them logically. There is none of the incoherent verbiage so typical of much alchemical literature, and though the reasons Geber advances in order to refute sceptics are of the usual kind they are expressed concisely and with restraint. Similar remarks apply to his descriptions of substances used in alchemy. Thus of mercury he says:

Argentvive, which is also called *Mercury* by the *Ancients*, is a viscous *Water* in the *Bowels* of the *Earth*, by most temperate *Heat* united, in a total *Union* through its least parts, with the substance of white subtile *Earth*, until the *Humid* be contempered by the *Dry*, and the *Dry* by the *Humid*, equally. Therefore it easily runs upon a plain *Superficies*, by reason of its *Watery Humidity*; but it adheres not although it hath a viscous *Humidity*, by reason of the *Dryness* of that which contemperates it, and permits it not to adhere. It is also (as some say) the *Matter* of *Metals* with *Sulphur*. And it easily adheres to three *Metals*, *viz.* to *Saturn*, and *Jupiter*, and *Sol*, but to *Luna* more difficultly. To *Venus* more difficultly than to *Luna*; but to *Mars* in no wise, unless by *Artifice*. Therefore hence you may collect a very great *Secret*. For it is amicable, and pleasing to *Metals*, and the *Medium* of conjoyning *Tinctures*; and nothing is submerged in *Argentvive*, unless it be *Sol*. Yet *Jupiter* and *Saturn*, *Luna* and *Venus*, are dissolved by it, and mixed; and without it, none of the *Metals* can be gilded. It is fixed, and it is a *Tincture* of *Redness* of most exuberant *Reflection*, and fulgid *Splendor*; and then it recedes not from the *Commixtion*, until it is in its own *Nature*.

In this paragraph Geber lucidly explains his views on the constitution of mercury and why, although liquid, it does not wet glass or earthenware vessels. He also correctly describes the ease or difficulty with which amalgams may be made of mercury with lead, tin, gold, silver, and copper; adding that an amalgam of iron cannot be prepared directly. In point of fact it is by no means a simple matter to amalgamate iron, and it would be interesting to know whether Geber succeeded and if so what his 'artifice' was. The last sentence means that, when mercury is converted into a solid, the product is of a shining colour, so that the reference is probably to mercuric oxide, the particles of which are red and sparkling, or to cinnabar (mercuric sulphide).

Of gold, Geber says that it is

a *Metallick Body*, *Citrine*, ponderous, mute, fulgid, equally digested in the *Bowels* of the *Earth*, and very long washed with *Mineral Water*; under the *Hammer* extensible, fusible, and sustaining the Tryal of the *Cupel* ... According to this Definition, you may conclude, that nothing is true *Gold*, unless it hath all the *Causes* and *Differences* of the Definition of *Gold*. Yet, whatsoever *Metal* is radically Citrine, and brings to *Equality*, and cleanseth, it makes *Gold* of every kind of *Metals*. Therefore, we consider by the *Work* of *Nature*, and discern, that *Copper* may be changed into *Gold* by *Artifice*. For we see in *Copper Mines*, a certain *Water* which flows out, and carries with it thin *Scales* of *Copper*, which (by a continual and long continued Course) it washeth and cleanseth. But after such *Water* ceaseth to flow, we find these thin *Scales* with the dry *Sand*, in three years time to be digested with the *Heat* of the *Sun*; and among these *Scales* the purest *Gold* is found. Therefore, we judge, those *Scales* were cleansed by the benefit of the *Water*, but were equally digested by the heat of the *Sun*, in the *Dryness* of the *Sand*, and so brought to *Equality*. Wherefore, imitating *Nature*, as far as we can, we likewise alter; yet in this we cannot follow *Nature*.

Also *Gold* is of *Metals* the most precious, and it is the *Tincture* of *Redness*; because it tingeth and transforms every *Body*. It is calcined and dissolved without profit, and is a *Medicine* rejoycing, and conserving the *Body* in *Youth*. It is most easily broken with *Mercury*, and by the *Odour* [vapour] of *Lead*. ... Likewise *Spirits* are commixed with it, and by it fixed, but not without very great *Ingenuity*, which comes not to an *Artificer* of a stiff neck.

In this passage Geber describes one of the observations that led alchemists to believe in the natural transmutation of metals. Many copper ores contain small quantities of gold, but so little as to be overlooked in the normal smelting-process. Water flowing out of a copper-mine might well carry particles of gold among the sludge, and after the detritus had weathered the scintillating specks of gold could attract attention in bright sunshine. The alchemists did not realize that the gold was present from the start; instead they drew the false conclusion that the 'mineral water' and the heat of the Sun had converted copper into the precious metal.

Geber is equally clear in his descriptions of various chemical operations. His directions for purifying salt, for instance, might have been taken from a modern laboratory manual:

Common salt is cleansed thus. First burn it [heat it strongly], and cast it combust into hot water to be dissolved; filter the solution, which congeal [crystallize] by gentle fire. Calcine the congelate for a day and a night in moderate fire, and keep it for use.

On sublimation, Geber says that it was invented for the purification of such substances as sulphur, arsenic [sulphides], and tutia [zinc oxide].

We were constrained to cleanse these from their burning *Unctuo-*

sity, and from the *Earthy Superfluity*, which they all have. And this We could effect by no *Magistery*, but by *Sublimation* only. ... *Sublimation* is the *Elevation* of a dry Thing by *Fire*, with adherency to its Vessel. But *Sublimation* is diversely made, according to the *Diversity* of *Spirits* to be sublimed ... whence it is necessary that the *Artificer* should apply to his *Sublimation* a three-fold *Degree* of *Fire*: one proportionate in such wise, that by it may ascend only the *Altered*, and more *Clean*, and more *Lucid*; until by this he manifestly see, that they are cleansed from their *Earthy Feculency*. The other *Degree* is, that what is of the pure *Essence* of them remaining in the *Feces*, may be sublimed with greater force of *Fire*, *viz.* with *Ignition* of the *Bottom* of the *Vessel*, and of the *Feces* therein, which may be seen with the *Eye*. The third *Degree* of *Fire* is, that unto the *Sublimate* without the *Feces*, a most weak *Fire* be administered, so that scarcely any thing of it may ascend, but only that which is the most subtile part thereof, and which in our *Work* is of no value.

Such passages indicate not only that Geber was widely versed in practical chemical operations, but that he had an inquiring and systematic mind anxious to understand what was going on in those operations and the reasons for performing them. He was much nearer in spirit to a chemist than an alchemist, and fully appreciated the value of experiment. '*Son* of doctrine', he says, 'search out *Experiments*, and cease not; because in them you may find *Fruit* a *Thousand-fold*.' This enthusiasm for experiment accounts for the clarity of his descriptions: he was generally writing from first-hand acquaintance with the substances and manipulations and not drawing upon secondary sources. No one who had not himself performed the experiment could give such precise instructions as the following, for the preparation of nitric acid:

First take of vitriol of Cyprus [copper sulphate], one pound; of saltpetre, two pounds; and of Jamenous allom, one fourth part; extract the water with redness of the alembic (for it is very solutive) ... This is also made much more acute, if in it you shall dissolve a fourth part of sal ammoniac; because that dissolves gold, sulphur, and silver.

The 'solutive water' is *aqua fortis* or nitric acid, and the 'much more acute' liquid is *aqua regia*, a mixture of nitric and hydrochloric acids. *Aqua regia* does in fact dissolve gold and sulphur, but though it attacks silver the metal is not dissolved.

In general theory, as we have seen, Geber accepts the assumption that metals are composed of sulphur and mercury, using those words to signify ideal or philosophic substances to which ordinary sulphur and mercury are only very crude approximations. Like most other alchemists of the time, he assigns the preponderating role to mercury; the nobler the metal the greater the proportion of mercury in it. Thus gold is of the most 'subtile substance of argentvive' with a small amount of the substance of sulphur, 'clean, and of pure redness, fixed, clear, and changed from its own nature', which tinges the mercury to golden yellowness.

More interesting than these alchemical commonplaces is Geber's explanation of how he believes the philosophers' stone would act, and what qualities must therefore be postulated in it. He says that the Stone must have the properties of oleaginy or oiliness, tenuity of matter, affinity, radical humidity, clearness of purity, a fixing earth, and tincture, and then proceeds to a description of the successive functions performed by the Stone in virtue of these properties. The first thing that is necessary after the projection of the Stone is its sudden and easy fusion, and this occurs because of its oleaginy. Next, the tenuity of the Stone makes it a very thin liquid when fused, so that it can immediately penetrate throughout the whole of the material to be transmuted. Affinity is necessary between the Stone and the material, otherwise the two would not adhere and cohere, while the radical humidity congeals and consolidates the similar parts of the material inseparably and for ever. The clearness of purity gives 'evident splendour', and at this stage remaining dross can be burnt away in the fire.

The sixth *Property*, is a *Fixing Earth*, temperate, thin, subtile, fixed, and incombustible, giving permanency of *Fixation* in the *Solution* of the Body adhering with it, standing and persevering against the force of *Fire*: for immediately after *Purification, Fixation* is necessary, and opportune.

Finally, the seventh property

is *Tincture*, giving a splendid and perfect *Colour*, *White*, or intensely *Citrine*, and *Lunification*, or *Solification* of *Bodies* to be transmuted; because after *Fixation*, a splendid *Tincture*, and *Colour* tinging another *Body*, or a *Tincture* colouring the *Matter* convertible into true *Silver*, or *Gold* (with all its certain and known differences) is absolutely necessary.

Most alchemists, while ready to praise the wonders of the Stone, are chary of advancing ideas of how it is to affect these wonders; but Geber had manifestly pondered the matter long and deeply. The scheme he elaborated indicates that he was working to a definite plan with a definite aim; the fact that he was attempting the impossible owing to insufficient knowledge and to fundamentally incorrect presuppositions should not make us overlook the penetration of his intelligence. He was essentially a forerunner of those modern chemists who set out to synthesize a substance that shall have certain pre-selected properties, and it is not difficult to believe that, had he lived in the nineteenth century or later, he might have emulated the achievements of Emil Fischer or Sir William Perkin. As Richard Russell remarks in the preface to his translation of Geber's works: 'The *Eminency* and *Worth* of this *Author* need no *Apology*; his *Works* sufficiently commend him.'

THE NEW PEARL OF GREAT PRICE

An alchemical work highly esteemed by the alchemists themselves was 'The New Pearl of Great Price' (*Pretiosa Margarita Novella*), written in 1330 or thereabouts by Petrus Bonus. The author cannot be identified with certainty, but he may have belonged to the Avogadrus family of Ferrara; the book itself was written at Pola, then a city of the Italian province of Istria but now, under the name of Pulj, included in Yugoslavia. It was first edited in 1546 by Janus Lacinius Therapus in an abridged and paraphrased form, and a further abridged edition in English was published by A. E. Waite in 1894.

A remarkable feature of the book is that Bonus, though declaring that the whole secret of transmutation can be learned in a single day, or even in a single hour, admits at the end that he himself had never been successful in the Art. 'This unusual candour in an alchemical writer', says Lynn Thorndike pertinently, 'rather disarms our criticism, and makes us feel that we have to do with a genuinely first-hand document which reflects the relation of alchemy to the thinking world of a particular past period rather than with the forgery of some quack or romancer who directed his appeal to gullible and unthinking followers of a current fad and delusion.'

Although Bonus asserted that essential alchemical knowledge could be transmitted in a very short time, he goes on to explain that the search for that knowledge is very difficult, partly because the adepts use words not only in their ordinary sense but in allegorical, metaphorical, enigmatical, equivocal, and even ironical ways. Moreover, alchemical writers often contradict one another, and working alchemists use different practical methods.

In the manner of the Schoolmen, Bonus first marshals the argument that can be advanced against the truth of alchemy, and does so very cogently; then, later in the book, he shows how these arguments may be refuted. Some of the reasons militating against the reality of the Art are as follows. The metals are composite substances, but the alchemists do not know their exact composition, therefore cannot produce them. Neither are the peculiar manner and mode of metallic composition known. In the production of metals, Nature uses a mixed heat, derived partly from the Sun and partly from the centre of the Earth; this cannot be imitated by alchemists. In Nature, the generation of metals takes thousands of years, and occurs in the bowels of the Earth. This process cannot be hastened appreciably by heat, because excessive heat would hinder development, and neither can glass or earthenware vessels replace the natural womb of metals in the ground. Alchemy cannot produce animal life, yet animals are easily decomposed and putrefied; much less therefore can it produce metals, which are of a much stronger composition. It is true that metals are generically alike, but they are different specifically; now it is impossible to change one species into another. Metals are formed under the movements and influences of the stars; but these movements no human mind can fix or direct to any given spot. It is easier to destroy things than to make them; but we can hardly destroy gold, so how should we make it? The ancient philosophers were in the habit of teaching all the arts and sciences they knew to their disciples, and of writing them in their books; but of alchemy they mention never a word. The Stone is supposed to harden lead and tin into gold, and to soften silver, iron, and copper into gold; but it is impossible that one and the same thing should produce opposite effects. It is not correct to call metals other than gold and silver imperfect, since in their own way they are complete. Critics also say that alchemists can merely alter and not transmute, and that alchemical gold and silver are not the same as those precious metals naturally derived.

This would seem a powerful case, but Bonus thinks it may be easily refuted. He does not consider it necessary to go back to ultimate constituents, for the proximate constituents of all metals are known to be sulphur and mercury; the base metals are imperfect and diseased but were ordained by Nature to become gold eventually and are already well on the way. It therefore only remains for the alchemist to cure the diseased metals by ridding them of superfluity of sulphur. Astrological influences need not be taken into account, since the celestial power is a constant factor in the growth or transmutation of metals; there is no virtue in working under a particular constellation or sign. As to the length of time required, Bonus is of the opinion that, to have reached even the base-metal stage, the process of metallic generation must have been going on so long that the final stage may be accomplished very quickly – in fact instantaneously on the addition of the Stone. Transmutation is merely the work of Nature aided by the Art and directed by the divine will. The inconsistencies that critics find in the Stone cannot be explained on natural grounds but must be accepted by faith, which has no difficulty in accepting the Christian miracles. In alchemy, work without faith is foredoomed to failure; the Art is a divine secret, and this explains also why it was not revealed by the ancient philosophers.

Descending to detail, Bonus says that the principles of alchemy are twofold, natural and artificial. The natural principles are the causes of the four elements, of the metals, and of all that belongs to them. The artificial principles are sublimation, separation, distillation, calcination, coagulation, fixation, and ceration, besides all the tests, signs, and colours by which the artificer can tell whether these operations have been properly performed or not. Thus the tests to be withstood by gold are incineration, ignition, fusion, and exposure to corrosive vapours: and in this connexion Bonus remarks that to identify a substance we must observe its properties. Every metal that exhibits all the qualities of gold – orange colour, fusibility, malleability, indestructibility, homogeneous nature, and great density – must be regarded as gold, and it would be mere sophistry to argue that it was not. The opponents of alchemy must therefore discover some property in which alchemical gold differs from natural gold if they are to maintain their position.

Unfortunately, Bonus is not always as rational as the foregoing passage might indicate. When he turns to vindicating alchemy by analogy he makes the most reckless statements in a completely uncritical spirit.

Something closely analogous to the generation of alchemy [he says] is observed in the animal, vegetable, mineral, and elementary world. Nature generates frogs in the clouds, or by means of putrefaction in dust moistened with rain, by the ultimate disposition of kindred substances. Avicenna tells us that a calf was generated in the clouds, amid thunder, and reached the earth in a stupefied condition. The decomposition of a basilisk generates scorpions. In the dead body of a calf are generated bees, wasps in the carcase of an ass, beetles in the flesh of a horse, and locusts in that of a mule. ... The same law holds good in the mineral world, though not to quite so great an extent.

The point of this analogy is that by the fortuitous meeting of the constituents of an object, the object itself may result, though formed in a way different from the normal one. Hence although the normal development of the metallic substance into gold is very slow as a natural process, it does not follow that that process is the only one. But the best analogy to the philosophers' stone is furnished by a consideration of smoke. This may become fixed or condensed as soot, and

1. Alchemical laboratory, by Teniers

2. Sixteenth-century German alchemical laboratory

3. Seventeenth-century alchemical laboratory

4. An alchemist, Hennig Brand, discovers phosphorus
(after the painting by Joseph Wright)

5. 'The pot to-breketh, and farewel! al is go!' (See p. 181)

6. The Alchemist, by Cornelius Bega

7. Four alchemical authorities, Geber, Arnold of Villanova, Razi, and Hermes, presiding over operations in a laboratory. (From Thomas Norton, *Ordinall of Alchimy*, London, 1652)

8. Thomas Norton in his laboratory. Note the balance in a glass case, the symbols for gold and silver on the table, and the stills in the foreground. (From Thomas Norton, *Ordinall of Alchimy*, London, 1652)

9. Astrological diagrams concerning alchemy. See p. 197.
(From Thomas Norton, *Ordinall of Alchimy*, London, 1652)

10. Alchemical furnaces in use for digestion and multiple distillation.
(From Thomas Norton, *Ordinall of Alchimy*, London, 1652)

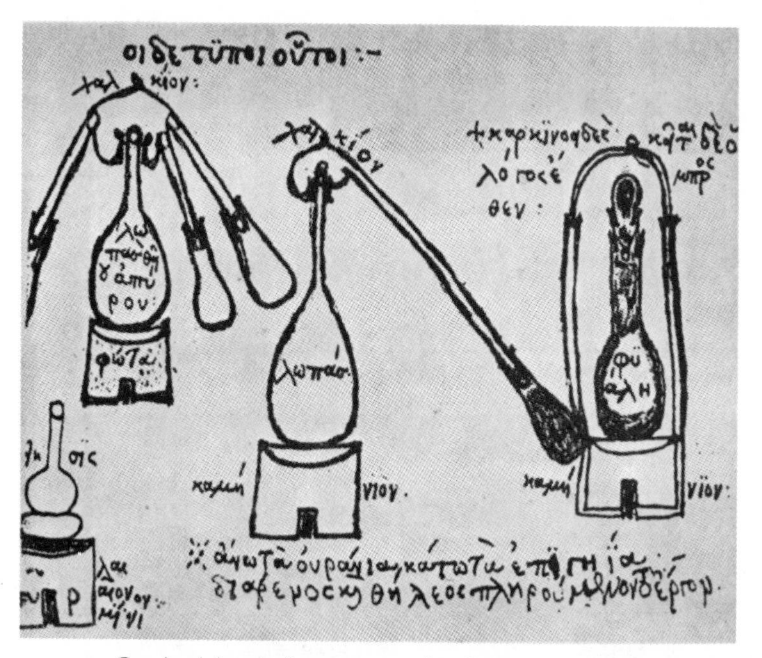

11. Greek alchemical apparatus for distillation (alembics) and digestion. See p. 48. (From Berthelot, *Collection des Anciens alchimistes grecs*)

12. Frontispiece of Béroalde de Verville's *Le Songe de Poliphile* (Paris, 1600). Note the alchemical symbolism. Success is indicated by the birth of the phoenix

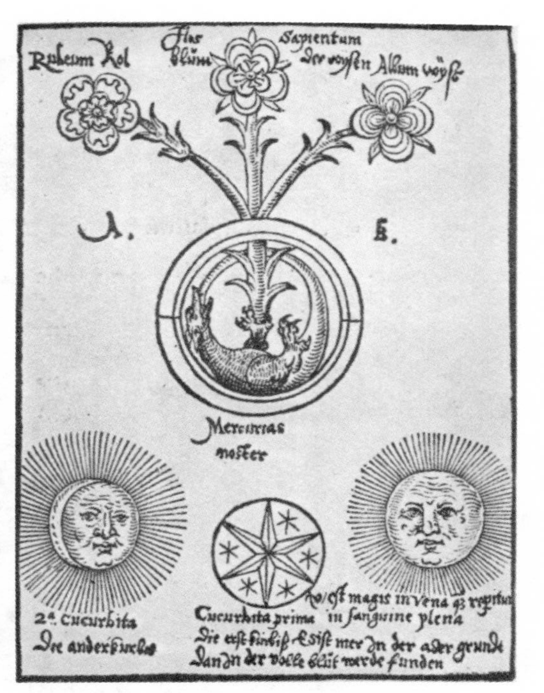

13. Ouroboros. See p. 159. (From H. Reusner, *Pandora: das ist, die edelst Gab Gottes*, Basel, 1582)

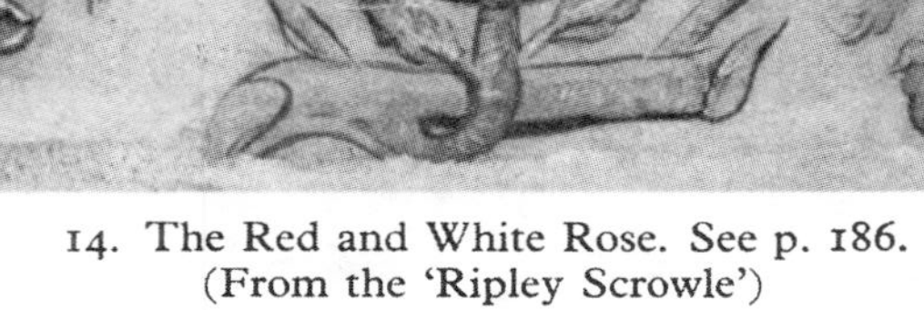

14. The Red and White Rose. See p. 186.
(From the 'Ripley Scrowle')

15. The Alchemical Process in the Zodiac.
(From the 'Ripley Scrowle')

16. The Mountain of the Adepts. (From S. Michelspacher, *Cabala, Speculum Artis et Naturae, in Alchymia*, Augsburg, 1654)

17. Alchemists at work. (From the *Liber Mutus*)

18. Four Stages of the Alchemical Process. (From J. D. Mylius, *Philosophia Reformata*, Frankfort, 1622)

19. Sulphur and Mercury, Sun and Moon. (From J. C. Barchusen, *Elementa Chemiae*, Leiden, 1718)

20. Alchemical library and laboratory. (From M. Maier, *Tripus Aureus*, Frankfort, 1678)

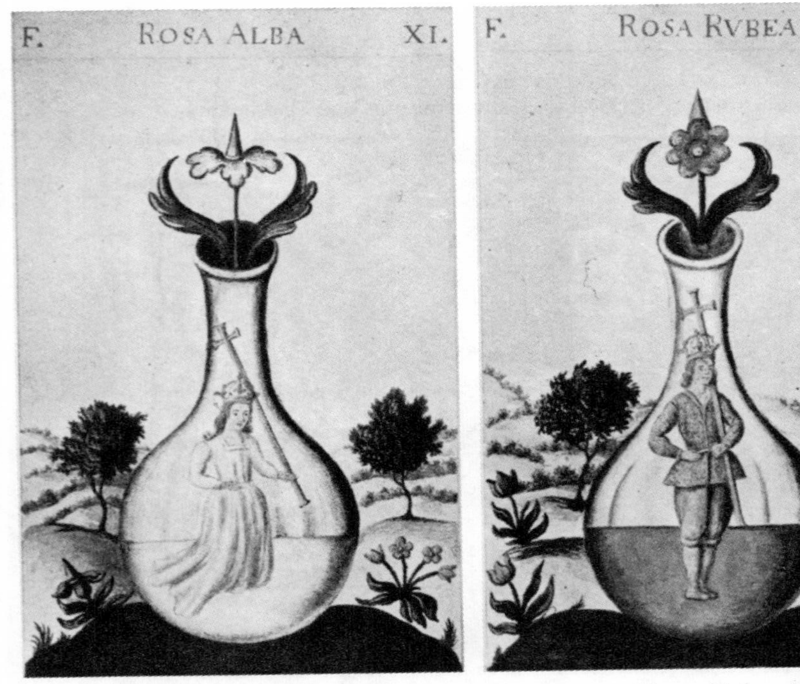

21, a and b. The White and Red Roses. See p. 242. (From *Trésor des Trésors*, MS 975, Bibliothèque de l'Arsenal; photo: Bibliothèque Nationale)

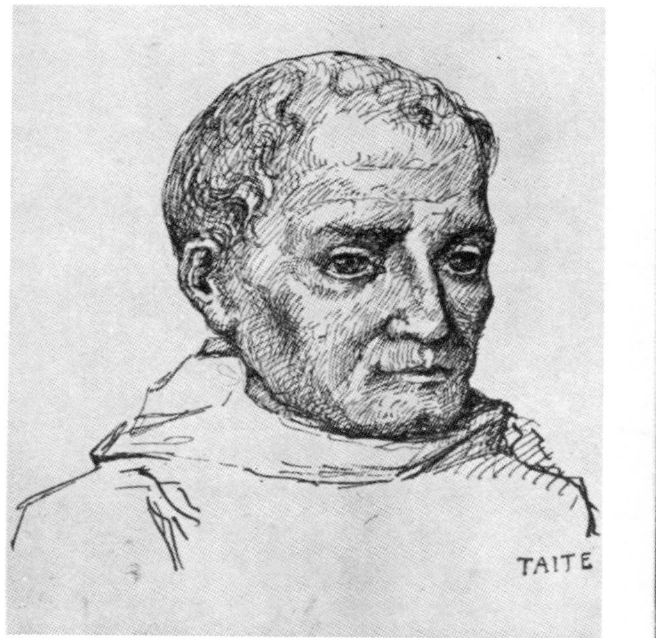

22. Portrait of Nicolas Flamel. See p. 246. (From A. Poisson, *Nicolas Flamel, sa vie, ses fondations, ses oeuvres*, by courtesy of the Bibliothèque Chacornac, Paris)

23. The church of Saint-Jacques-la-Boucherie. See p. 239. (From A. Poisson, *Nicolas Flamel, sa vie, ses fondations, ses oeuvres*, by courtesy of the Bibliothèque Chacornac, Paris)

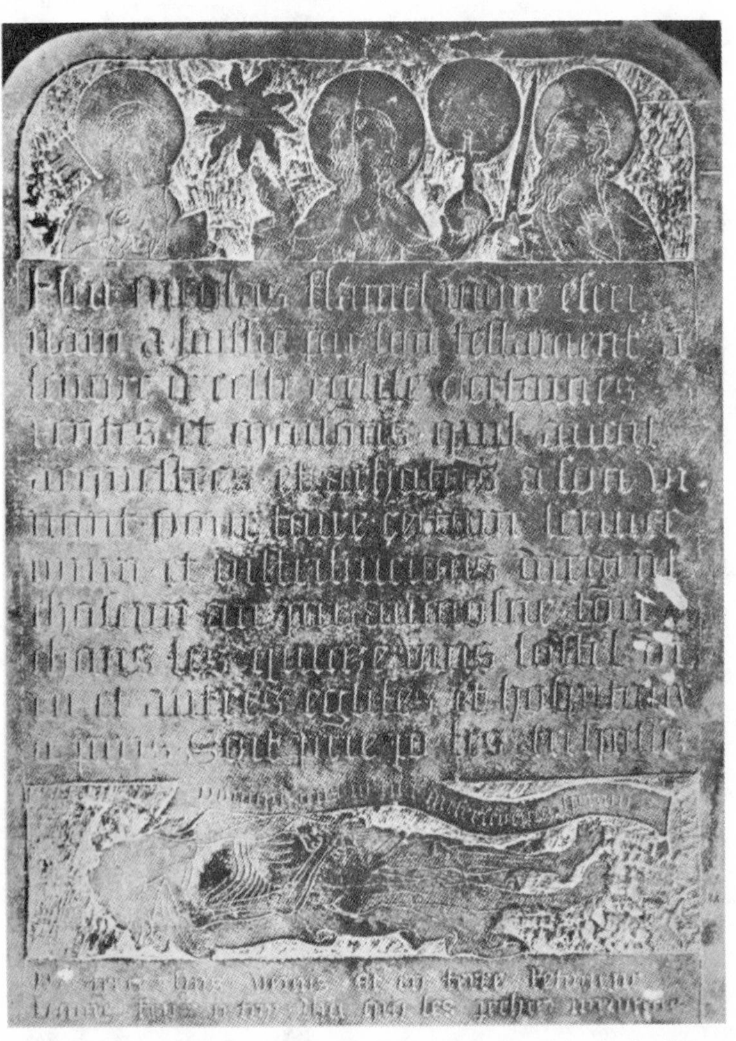

24. The Tombstone of Nicolas Flamel, now in the Musée de Cluny. (Photo: Archives Photographiques)

25. Ivory Alchemical Statuettes of the Sun and Moon. (By courtesy of the Science Museum and the executors of the late Dr Sherwood Taylor)

26. The Sulphur-Mercury Theory. See pp. 74, 75. (From M. Maier, *Symbola Aureae Mensae Duodecim Nationum*, Frankfort, 1617)

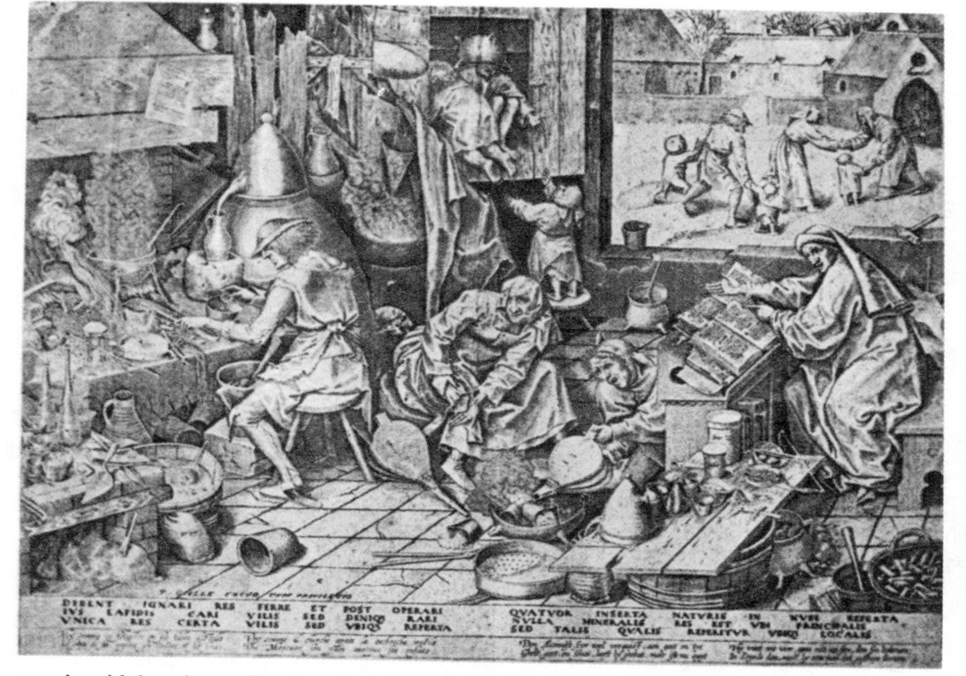

27. An Alchemist at Work, after Brueghel. (An original print of the engraving after Brueghel, in the possession of Professor John Read; by courtesy of Professor Read and of Thomas Nelson Ltd)

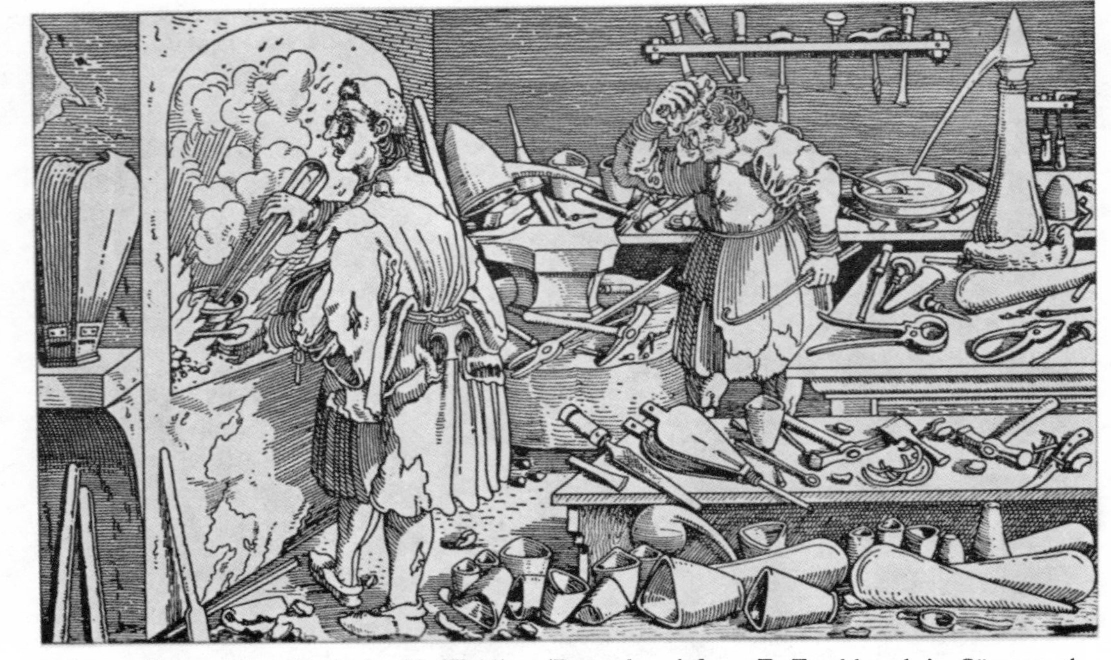

28. An Alchemist at Work, by H. Weiditz. (Reproduced from F. Ferchl and A. Süssenguth, *A Pictorial History of Chemistry*, by courtesy of William Heinemann Ltd)

29. Figures d'Abraham Juif. See p. 241. (Reproduced from J. Read, *Prelude to Chemistry*, by courtesy of G. Bell & Sons Ltd)

30. The Seventh Parable. (From Solomon Trismosin, *Splendor Solis*, Harley MS 3469; by courtesy of the British Museum)

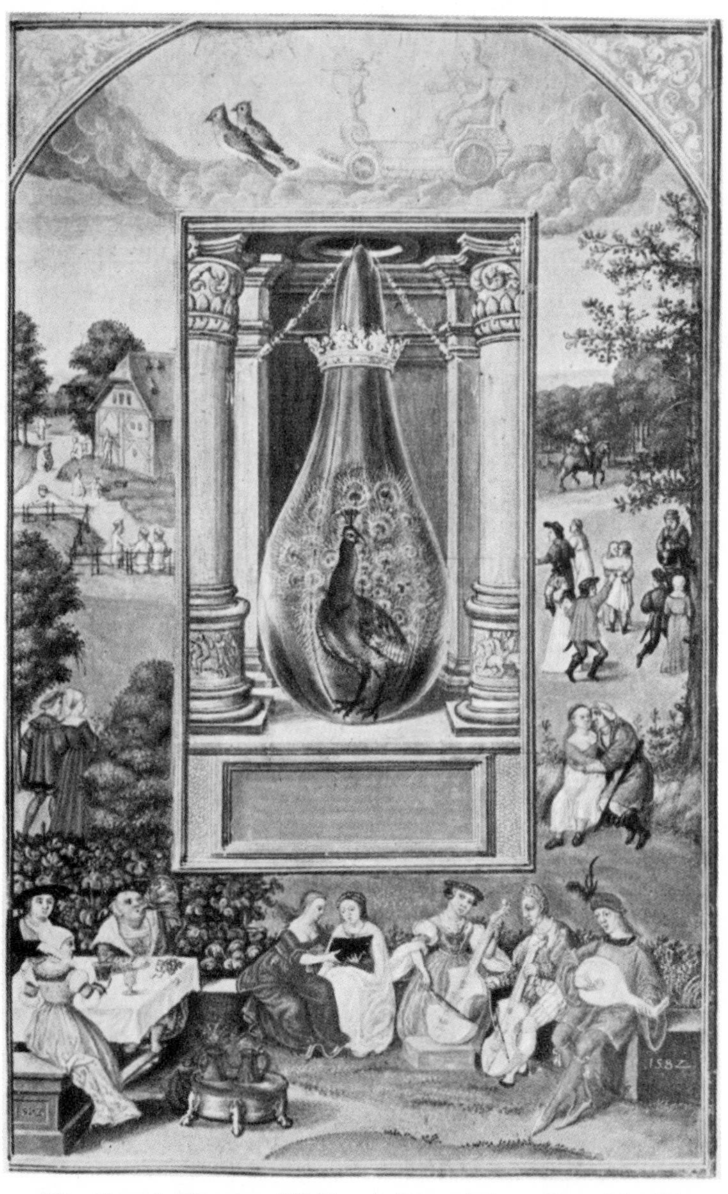

31. The Fourth Treatise, Fifthly. A Peacock in a Cucurbit. (From Solomon Trismosin, *Splendor Solis*, Harley MS 3469; by courtesy of the British Museum)

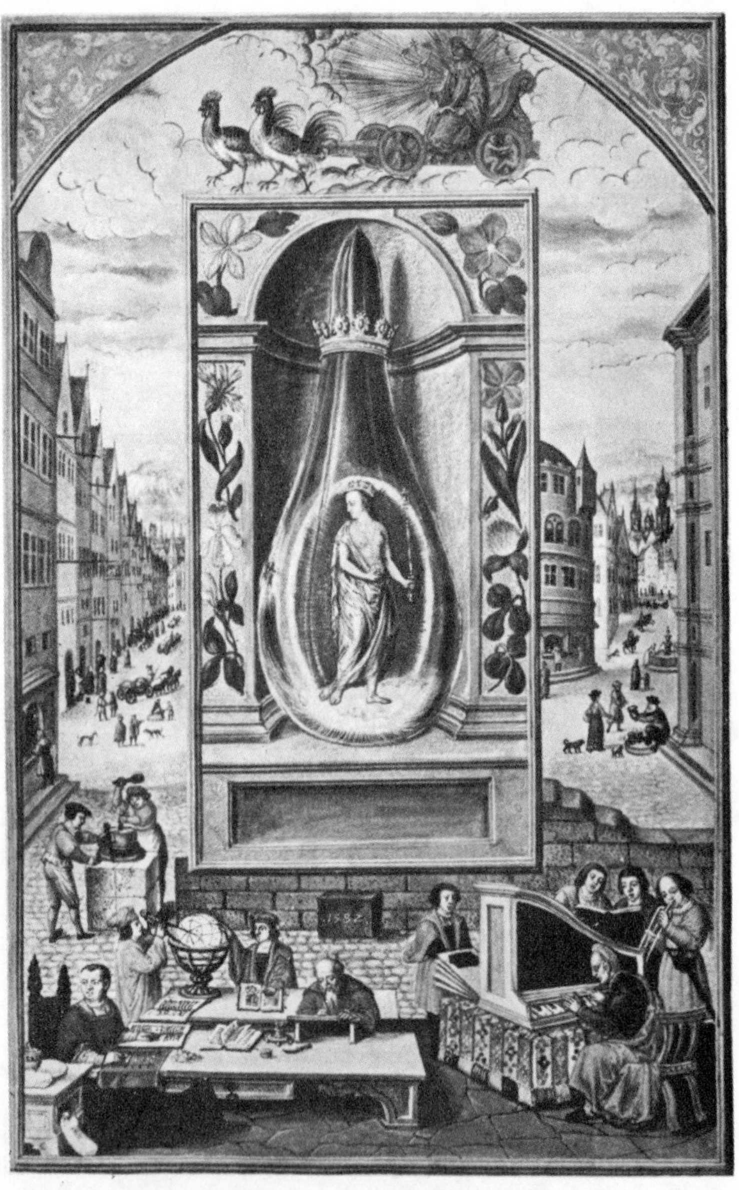

32. The Fourth Treatise, Sixthly. A Queen in a Cucurbit. (From Solomon Trismosin, *Splendor Solis*, Harley MS 3469; by courtesy of the British Museum)

33. Symbols of calcination, distillation, coagulation, and solution, and the extraction of the philosophic Mercury from the *prima materia* by means of the philosophic fire. See p. 162. (From the Codex Germanicus 598, by courtesy of the Bayerische Staatsbibliothek, Munich)

34. The Grand Hermetic Androgyne trampling underfoot the four elements of the *prima materia*. (From the Codex Germanicus 598, by courtesy of the Bayerische Staatsbibliothek, Munich)

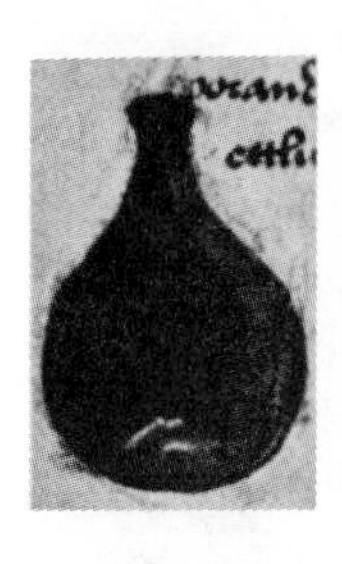

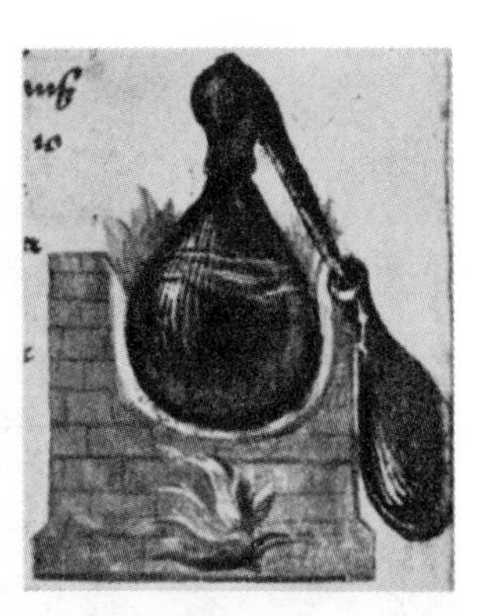

35. A flask, a pelican, a double pelican, an alembic in an air-bath, and symbols of the four evangelists. (From the Codex Germanicus 598, by courtesy of the Bayerische Staatsbibliothek, Munich)

36. The Alchemist, by J. Stradanus. (From the Palazzo Vecchio, Florence; photo: Alinari)

here a spirit, as it were, has evaporated from the fire and has assumed a corporeal form. Similarly, out of dry vapours are generated dry things, and from humid vapours come moist substances; the digestion and proportionate commixture of both these kinds produce the diversity of all generated species, according to the exigency of their nature. And as these vapours, whether dry or moist, are actively fugitive and ascending, so they are potentially permanent and resting. If the alchemist by the preparation of this proper matter in a proper vessel, with the suitable degree of fire, paying due attention to the significance of the sequence of colours, can obtain that which constitutes the essence of gold in a concentrated volatile or spiritual form, he can pervade with it every particle of a base metal, and thus transmute it into gold. This action of foreign bodies anyone can observe on the surface of metals; tutia [zinc oxide] imparts to copper the colour of gold, orpiment and arsenic colour it white like silver; the fumes of Saturn congeal quick-silver; the rind of the pomegranate converts iron into steel, and the fumes of burnt hair give to silver an orange tinge. Let us suppose the metals to be penetrated by some more powerful and all-pervading agent in their very inmost parts, and we have something closely resembling the action of the philosophers' stone. The spirits of metals potentially contain their bodies, and this potentiality may at any moment become actual, if the artist understands, and knows how to imitate, Nature's method of working.

In the composition of metals, Bonus emphasizes the supreme importance of mercury, which contains within itself a subtle intrinsic sulphur. All metals except gold contain an extrinsic sulphur as well, which is the cause of their imperfection. Thus (following Waite's translation, slightly modified):

It should be observed that there are in all varieties of metal, except gold, two kinds of sulphur, one external and burning, the other inward and non-combustive, being of the substantial composition of quicksilver. The outward sulphur is separable from these; the inward sulphur is not. The outward sulphur, then, is not actually united to the quicksilver; hence the quicksilver cannot be really burnt by it. Therefore when the quicksilver is purified by the removal of the outward sulphur, it is restored to its original condition, and can be transmuted into gold or silver, whether it be found in tin and lead, or in iron and copper; and we may justly conclude from these considerations that when the philosophers' stone is projected upon iron or copper in a liquefied state, it mingles in a moment of time with all the particles of quicksilver existing in them, and with these only, and perfects them into the purest gold, while all the particles of external sulphur are purged off, because they are not of a nature homogeneous with that of the Stone. For quicksilver always most readily combines with any substance that is of the same nature as itself, and rejects and casts out everything heterogeneous. It does not matter what are the other constituent parts of a metal; if it be a metal it must contain quicksilver, and that quicksilver can be changed into gold by means of the philosophers' stone. So we see that, in the case of milk, the coagulant clots only those parts of the milk that are of a nature homogeneous with its own.

To prove that mercury is not essentially affected by extrinsic sulphur, Bonus reminds his readers that though when mercury and sulphur are heated together cinnabar is formed, yet on further heating the mercury is regained pure and clean. The formation of the cinnabar was thus merely a partial combustion of the sulphur; the mercury had undergone no real change.

When alchemical authors have given their views on the method of obtaining the philosophers' stone, they are usually content to say that at this stage projection may be effected and to leave the matter there. Bonus, however, refers at some length to the 'ferment' – a word which, as he says, they seem to use in two senses, meaning either the Stone itself, or that which perfects and completes the Stone. The idea of a ferment in connexion with transmutation is very frequently encountered, and since Bonus writes more clearly than most alchemists it is worth while listening to his explanation.

In the first sense our Stone is the leaven of all other metals, and changes them into its own nature – a small piece of leaven leavening a whole lump. As leaven, though of the same nature with dough, cannot raise it until, from being dough, it has received a new quality which it did not possess before, so our Stone cannot change metals until it is changed itself, and has added to it a certain virtue which it did not possess before. It cannot change, or colour, unless it has first itself been changed and coloured. Ordinary leaven receives its fermenting power through the digestive virtue of gentle and hidden heat; and so our Stone is rendered capable of fermenting, converting, and altering metals by means of a certain digestive heat, which brings out its potential and latent properties, seeing that without heat neither digestion nor operation is possible. The difference between ordinary leaven and our ferment is that common leaven loses nothing of its substance in the digestive process, while digestion removes from our ferment all that is superfluous, impure, and corruptive, as is done by Nature in the preparation of gold. It is because our ferment assimilates all metals to itself, just as common leaven assimilates to itself the whole mass of dough, that it has received this name from the sages. ... The action of our ferment is not quiet analogous, however, to that of leaven. For leaven changes the whole of the dough into a kind of leaven; but our Stone, instead of converting metals into the Tincture, transmutes them only into gold.

This is because, in spite of its great power, the Stone loses some of its virtue in effecting transmutation, so cannot change metals to as high a state as its own: 'the virtue of our Stone is great, yet, on being mixed with common metals, its action is slightly affected by their impurity, and does not change them quite into its own likeness, but only into gold'.

In the other sense of the word, the ferment is invisible to the eye, but capable of being apprehended by the mind.

It is the body which retains the soul, and the soul can shew its power only when it is united to the body. Therefore when the artist sees the white soul arise, he should join it to its body in the same instant, for no soul can be retained without its body. This union takes place through the mediation of the spirit, for the soul cannot abide in the body except through the spirit, which gives permanence to their union, and this conjunction is the end of the work. Now, the body is nothing new or foreign; but that which was before hidden becomes manifest and that which was manifest becomes hidden. The body is stronger than soul and spirit, and if they are to be retained it must be by means of the body. The body is the form, and the ferment, and the Tincture of which the sages are in search. It is white actually and red potentially; while it is white it is still imperfect, but it is perfected when it becomes red.

This passage is not by any means pellucid, but perhaps it means that, in the second sense of the word ferment, the reference is to a critical operation to be conducted at a precise moment immediately before the preparation of the Stone is completed. The alchemists seem, or at least Bonus in the 'New Pearl' seems, to have envisaged the Stone as a very unstable substance until perfected, so that negligence in the closing stages might ruin everything. Whether this is the intended meaning or not would be difficult to decide, but clearly there were rapid changes to be expected at this point, particularly changes of colour.

> Dealbation must come first, for it is the beginning of the whole work, and then the rubefaction must follow, which is the perfection of the work. Since the entire substance – the soul united to the body by the spirit – is of the pure nature of gold, it follows that whatever it converts it must convert into gold. At first, indeed, the whole mass is white because quicksilver predominates; but the gold is dominant, though hidden, in it; when it is ferment, the mass in the second stage of the magistery becomes red in the fullness of the potential sense, while in the third stage, or the second and last decoction, the ferment is actively dominant, and the red colour becomes manifest, and possesses the whole substance. Again, we say that this ferment is that strong substance which then turns everything into its own nature. Our ferment is of the same substance as gold; gold is of quicksilver, and our design is to produce gold.

Bonus relied for his authorities on classical and Muslim sources and no medieval Latin alchemists are mentioned by name in the 'New Pearl'. He particularly wished to bring alchemical theory into line with Aristotle's views, and suggested that the Stagirite's arguments against alchemy in the *Meteorologica* were youthful opinions corrected in later life. In the Middle Ages Aristotle's infallibility was almost entirely unquestioned; he had been accepted as semi-official philosopher by both Islam and the Church, and on disputed questions like those of alchemy great efforts were always made by each side of the dispute to show that Aristotle had supported its contentions. But Bonus was not content with Aristotle; he also claimed that Ovid's *Metamorphoses* dealt esoterically with the philosophers' stone, and that many other ancient poems and myths had hidden alchemical meanings.

JOHN DASTIN

John Dastin – the name is also found as Dastyn, Dausten, Daustein, and other variants – was the foremost alchemist of his time, but we know little of him except that he was a member of a religious order and lived with extreme frugality. His period can, however, be fixed more or less closely, for he wrote alchemical letters to John XXII, who was Pope from 1316 to 1334, and to Cardinal Orsini, who wore the red hat from 1288 to 1342. Dastin therefore lived in the first half of the fourteenth century. At that time the seat of the Papacy was at Avignon, and it seems that so much alchemical or counterfeit gold was being coined or otherwise circulated in France that John XXII felt compelled to issue a decretal against such practices. This decretal ran as follows, in a translation made by J. R. Partington:

Poor themselves, the alchemists promise riches which are not forthcoming; wise also in their own conceit, they fall into the ditch which they themselves have digged. For there is no doubt that the professors of this art of alchemy make fun of each other because, conscious of their own ignorance, they are surprised at those who say anything of this kind about themselves; when the truth sought does not come to them they fix on a day for their experiment and exhaust all their arts; then they dissimulate their failure so that finally, though there is no such thing in nature, they pretend to make genuine gold and silver by a sophistic transmutation; to such an extent does their damned and damnable temerity go that they stamp upon the base metal the characters of public money for believing eyes, and it is only in this way that they deceive the ignorant populace as to the alchemic fire of their furnace. Wishing to banish such practices for all time, we have determined by this formal edict that whoever shall make gold or silver of this kind or shall order it to be made, provided the attempt actually ensues, or whoever shall knowingly assist those actually engaged in such a process, or whoever shall knowingly make use of such gold or silver either by selling it or giving it in payment for debt, shall be compelled as a penalty to pay into the public treasury, to be used for the poor, as much by weight of genuine gold or silver as there may be of alchemical metal, provided it be proved lawfully that they have been guilty in any of the aforesaid ways; as for those who persist in making alchemical gold, or, as has been said, in using it knowingly, let them be branded with the mark of perpetual infamy. But if the means of the delinquents are insufficient for the payment of the amount stated then the good judgement of the justice may commute this penalty for some other (as for example imprisonment or another punishment, according to the nature of the case, the difference of individuals and other circumstances). Those, however, who in their regrettable folly go so far as not only to pass monies thus made but even despise the precepts of the natural law, overstep the limits of their art and violate the laws by deliberately coining or casting or causing others to coin or cast counterfeit money from alchemical gold or silver, we proclaim as coming under this animadversion, and their goods shall be confiscate, and they shall be considered as criminals. And if the delinquents are clerics, besides the aforeside penalties they shall be deprived of any benefices they shall hold and shall be declared incapable of holding any further benefices.

John Dastin took up the cudgels on the side of the alchemists, and wrote to the Pope explaining the truth about 'the most noble matter, which, according to the tradition of all philosophers, transforms any metallic body into very pure gold and silver' and which 'makes an old man young and drives out all sickness of the body'. Manuscripts of the letter are still preserved, and the Latin text has been translated into English by C. H. Josten. It is rather difficult to understand the details of Dastin's theories, but he seems to imply that the 'ferment' or 'sulphur' intended to convert mercury into gold is in fact nothing but highly purified gold itself, but that before the action can take place the mercury must be fixed or rendered non-volatile:

Gold is more valuable than all other metals, because it contains in itself the essence of any metal. It tinges them and vivifies them, because it is the ferment of the elixir, without which the philosophers' medicine can by no means be perfected, like as dough cannot be fermented without a ferment. It is indeed as leaven to dough, as curd to milk for cheese, and as the musk in good perfumes. ... For those two bodies duly prepared are mercury and very pure sulphur, because if mercury is properly coagulated it transmutes into genuine gold and silver. If the mercury has been pure, the force of white, non-burning sulphur will congeal it. And that sulphur is the best one that those who practise alchemy can find or receive so as to convert it into silver. If however the pure and very good sulphur be of a clear red colour, and if there be in that sulphur the force of a simple non-burning fieriness, it will be the best thing that alchemists can find so as to make gold. And again: The ferment of gold is gold, and the ferment of silver is silver, and there are no other [suitable] ferments on earth.

This theory is elaborated in the letters to Orsini, where Dastin indicates that the red sulphur he postulates in gold is sufficient to convert suitably prepared mercury into the Red Elixir. In this operation, the product is first white, and then, after further heating, it assumes a clear red colour. The process, Dastin assures us, is quite simple, and may be carried out over a gentle fire in a hermetically sealed glass vessel, through the wall of which the changes of colour may be watched. The fire should be so regulated that the vapours continually ascend and do not solidify at the top of the vessel. The whole operation takes about 100 days.

In another work ascribed to Dastin the same ideas are found, but more emphasis is laid upon the function of the mercury, which is the sperm and material of metals and the philosophers' stone, and from which the elixir can indeed be prepared without the addition of any other material substance. Dastin seems to endow the elixir with a spiritual nature, so that although it must be confined in some kind of matter, for otherwise it could not be manipulated, it nevertheless occupies no space: an idea to which we find an interesting parallel in the views of Paracelsus (p. 171). Sulphur, we may deduce, had so puzzled Dastin by its corrosive action on metals and by its inflammability that he tried to diminish its importance and finally banished it altogether. But in his suggestion that the elixirs could be made from just gold and mercury, or silver and mercury, he was paving the way for the popular belief in the 'multiplication' of metals, which some of his successors vehemently opposed.

As to the powers of the elixirs,

one part converts a million parts of any body you may choose into the most genuine gold and silver, according to which of the two elixirs was prepared. The red elixir has effective virtue over all other medicines of the philosophers to cure all infirmity, because, if it were an illness of one month, it cures it within one day; if it were an illness of a year, it cures it in twelve days. But if it were an inveterate illness, it cures it in a month. And therefore this medicine ought to be sought for by all, and before all other medicines of this world. This magistery is for kings and the great of this world, because he who possesses it has a never-failing treasure.

History does not record whether Dastin's exposition was successful in causing John XXII to revise his opinion of alchemy, but when the Pope died he left an enormous fortune, said to be of alchemical origin.

7
SIGNS, SYMBOLS, AND SECRET TERMS

MODERN sciences have their own systems of signs and symbols, and use technical terms unintelligible except to the initiated. In this respect they resemble alchemy, but there is one fundamental difference. For the most part, alchemical symbolism was allegorical, and though a particular symbolic story or drawing might convey more or less the same idea to different adepts, it generally had little relation to definitely ascertained facts. There was, it is true, some use of notation, such as in the adoption of planetary signs to indicate metals, but for the most part alchemical symbolism was an attempt to convey the philosophy of the subject in an esoteric way. Its purpose was thus quite dissimilar to that of modern scientific symbolism. From the chemical formula of a substance, the chemist can at once gather a large number of definite and experimentally verifiable facts about it, but it would be idle to try to apply alchemical symbols in the same way.

The systems of notation were merely a kind of shorthand, designed perhaps to save time more than to puzzle the vulgar, but unfortunately, except for the signs for the seven metals, there was little uniformity in the schemes adopted by different alchemists; even one and the same practitioner might use several distinct signs for a single substance or operation. Some of the signs frequently encountered are as follows:

○ or ☉ gold [Sun]
☽ silver [Moon]
♀ copper [Venus]
♂ iron [Mars]
☿ mercury
♄ lead [Saturn]
♃ tin [Jupiter]

🜍 sulphur
🜔 salt
🜄 water
🜂 fire
🜅 *aqua fortis*
🜃 earth
🜁 air

🜹 sal ammoniac
🝞 sublimation
🜐 sublimate of mercury
🜼 realgar
🜖 vitriol
🝭 retort
□ common salt

There would be no advantage in extending this list, but it is of interest to inquire into the origins of the signs for the metals, which are also the signs of the planets with which the metals were supposed to be in astrological connexion. The signs for the Sun and Moon, and thus for gold and silver, are obviously just representational, but those for the other planets and metals have been variously explained. Thus the sign for Venus (copper) was said to represent the goddess's looking-glass, that for Mars (iron) his shield and spear, and so on; but it seems equally likely, if not more so, that the signs are modifications of the initial letters of the Greek (or Greco-Egyptian, Greco-Babylonian) names of the planets. Thus the symbol of Venus might have come from the *Φ* of *Φωσφόρος* (*phosphoros*), that of Saturn from the *Κρ* of *Κρόνος* (*kronos*), and that of Mercury from the *Στ* of *Στίλβων* (*stilbon*).

The symbols for fire, air, water, and earth indicate the property of the first two to ascend and that of the second two to descend, while those for sublimation and retort are crudely pictorial. The symbol for sublimate of mercury (mercuric chloride) illustrates how signs could be coupled, and in conveying information concisely foreshadows much later practice. Coupling of signs goes back even to the times of the Alexandrian alchemists, who joined the sign for gold to that of silver in order to represent the gold-silver alloy known as electrum.

Not only the materials but also the operations or stages in the preparation of the philosophers' stone were sometimes represented by signs. Thus Pernety, in his *Dictionnaire Mytho-Hermétique*, gives the following correspondence between twelve steps in the Great Work and the signs of the zodiac:

1. Calcination	♈	Aries, the Ram
2. Congelation	♉	Taurus, the Bull
3. Fixation	♊	Gemini, the Twins
4. Solution	♋	Cancer, the Crab
5. Digestion	♌	Leo, the Lion
6. Distillation	♍	Virgo, the Virgin
7. Sublimation	♎	Libra, the Scales
8. Separation	♏	Scorpio, the Scorpion
9. Ceration	♐	Sagittarius, the Archer
10. Fermentation	♑	Capricornus, the Goat
11. Multiplication	♒	Aquarius, the Water-carrier
12. Projection	♓	Pisces, the Fishes

Hieroglyphics of this kind are often useful in guiding us through the tortuous ways of alchemical literature, and would be more so if there were less inconsistency in their usage. It is when we come to the more cryptic symbolism that doubts and difficulties multiply. This symbolism is of two kinds, literary and graphic, and we may consider them in turn. Literary symbolism in its simplest form is seen in the inveterate habit of the adepts of applying fanciful names to the substances and apparatus used in alchemy, principally by metaphor or real or supposed analogy. Thus an oval or spherical glass vessel which could be hermetically sealed was commonly referred to as the 'philosophers' egg', and that not merely because of its shape but with a vague reference to the egg out of which the universe was hatched; so that in a particular context it may be hard to decide whether the author is describing a piece of apparatus or trying to convey a fragment of doctrine.

Sol or Sun for gold is intelligible enough, but when a passage relates that *Sol* was devoured by the green dragon it is only an acquaintance with alchemical symbolism that enables us to appreciate its significance: which is that *aqua regia* was used to dissolve gold and that in this case the gold probably contained copper which would colour the acid bluish-green. 'The black crow' sometimes means lead; 'the white eagle', sal ammoniac; 'the flower of the Sun', the red elixir; a king and his son and five servants would represent gold, mercury, and the other five metals of alchemy; 'the grey wolf' meant antimony; and 'celestial dew' signified mercury, or the mercurial principle, or even the spirit of nature. Base metals were compared to lepers (p. 162) and were not seldom regarded as being in this diseased state on account of their sinful nature. *Azoth*, as a name for mercury (especially the hypothetical 'mercury of the philosophers'), is a corruption of an Arabic name of the metal, *al-zāūq*, but was considered to have great occult power. This was because it contains the first and last letters of each of the Greek, Roman, Arabic, and Hebrew alphabets.

The vocabulary of alchemy contains hundreds and possibly thousand of such *Decknamen* or 'cover-names', as the Germans call them, and in the majority of cases it is impossible to give precision to their significance. When a recognizable operation is being described the substances employed can occasionally be deduced, but they were rarely pure and their varying properties when contaminated hinder identification, at the same time accounting in part for the many different names often applied by the alchemists to one and the same solid or liquid. There were, however, a good many alchemical writers of a practical turn of mind, captured by the fascination of chemical experimentation for its own sake, and although we have little information concerning the standards of purity they attained or attempted, it is certain that they were familiar with a very wide range of mineral, vegetable, and animal products. The complexity of nomenclature is thus explicable on the ground of complexity of material as well as by the adepts' weakness for allegory and their desire to keep their knowledge from the profane.

The use of cover-names was, however, the least part of the symbolism employed. The Great Work of alchemy was intimately bound up with the whole religious and philosophical background, and for many who practised it the transmutation of metals was symbolical of the transmutation of imperfect man into a state of perfection. Conversely, metallic transmutation could be brought about only by divine aid and by men of pure life. These two tenets reacted upon one another and are complexly interwoven in alchemical thought. The unity of the world and all things in it was an unshakable belief; there was thus nothing illogical in the combination of mystical theology with practical chemistry, however incongruous it may seem to us today. It is this combination that accounts for the extraordinary character of the bulk of serious alchemical literature: serious as opposed to the clearly fraudulent writings of charlatans and to those of men who derided what was in fact a philosophy of life. It is also this combination that often makes it very difficult to decide whether a particular work of symbolism is intended to convey any real chemical information, or whether it is to be taken as speculative throughout.

With this in mind, we may turn to some of the works in which the symbolism is of a literary kind – remembering that where processes are described they may bear no relation to actual ones, and that many theories are quite without foundation in fact. But first let us hear what qualifications are necessary for the worker.

The Byzantine Greek alchemist Archelaos (p. 31), in a poem 'Upon the Sacred Art', written early in the eighth century A.D., says (in C. A. Browne's translation):

> With inspiration from above take heart
> And strive with certain aim to reach the mark.
> The work which thou expectest to perform
> Will bring thee easily great joy and gain
> When soul and body thou dost beautify
> With chasteness, fasts and purity of mind,
> Avoiding life's distractions and, alone
> In prayerful service, giving praise to God,
> Entreating him with supplicating hands
> To grant thee grace and knowledge from above
> That thou, O mystic, may'st more quickly know
> How from one species to complete this work. . . .
> Thy body mortify by serving God:
> Thy soul let wing to look on godliness:
> So shalt thou never have at all the wish
> To do or think a thing that is not right.
> For strength of soul is manliness of mind,
> Sagacious reasoning and prudent thought.
> All passions purify and wash away
> The stain of carnal joys with streams of tears
> Which flood thy weeping eyes, revealing thus
> The pain and anguish of a contrite heart.
> Mind well Gehenna's fire and Judgement Day.
> So live that thou deservedly may'st see
> The shadowless and everlasting light.
> And from thy lips let tuneful praise ascend
> With choirs of angels unto God most high,
> Who rules above with wisdom, king of all,
> The Father with the Word and Holy Ghost,
> For all eternity and endless time
> Forever and forever more. Amen.

The same insistence on the good life is found in a fifteenth-century dialogue between an alchemist and his son who wishes to be taught the art:

... Son upon condicion I shall thee leare [teach]
So that thou wylt on the Sacrament sweare
That thou shalt never write it in scripture
Nor teach yt to no man except thou be sure
That he is a perfeit man to God and also full of chariti.
Doing alle waies good deede and that he be full of humilitie
And that you know him not in lowde words but alwaies soft and still
And alle so preve whether his life be good or yll
And alle this shal thou sweare and alle so make a vow
If thou wylt have thys Cunning [knowledge] of me now
And the same Oath on booke they make to thee
Ere thou them let any parte of thys scyence know or see.

Another alchemist of about the same period says that the adept should fully trust in God, lead a rightful life, subdue falsehood, be patient and not ambitious, and take part in no sinful strife. Even such a matter-of-fact alchemist as the Latin Geber (p. 134), after mentioning several secular qualities requisite in the aspirant, further admonishes him:

> Our Art is reserved in the Divine Will of God, and is given to, or withheld from, whom he will; who is glorious, sublime, and full of all Justice and Goodness. And perhaps, for the punishment of your sophistical work [work directed solely to material transformation], he denies you the Art, and lamentably thrusts you into the by-path of error, and from your error into perpetual infelicity and misery: because he is most miserable and unhappy, to whom (after the end of his work and labour) God denies the sight of Truth. For such a man is constituted in perpetual labour, beset with all misfortune and infelicity, loseth the consolation, joy, and delight of his whole time, and consumeth his life in grief without profit.

The worker should in short look upon the mere transmutation of base metals into gold as a secondary object, subsidiary to the transmutation of the soul. This emphasis is repeated by countless alchemical authors, and it is perhaps significant that, as Sherwood Taylor was the first to point out, alchemy was remarkably free from the taint of black magic, invocations of demons, necromancy, and other evil practices contemporaneous with it for practically the whole period during which it flourished. Exceptions to the rule can be found, but the moral tone of mystical alchemy is very high, in Greek, Muslim, and Christian writings alike.

The alchemists could find examples of transmutation in the Bible, such as the conversion of water into wine, and biblical imagery is frequently to be met with in symbolic alchemical works; but that and similar practices are used in both directions. That is to say, on the one hand the fact that apparently miraculous changes have scriptural or other authoritative support is taken to confirm the possibility of material transmutation, while on the other hand the operations of practical alchemy are interpreted as having a theological or mystical application. Classical and other mythology served alchemical purposes, often suffering considerable sea-changes, and the whole language of symbolic alchemy presents as colourful an imagery as would be difficult to match elsewhere. Without always attempting to elucidate any genuine chemical facts that may possibly lie beneath them, we may sample some of the vast literature of this *genre* that accumulated through the centuries. In some cases the mystical significance and even the physical basis may be fairly clear; in others, one man's guess is as good as his neighbour's. This is a passage, translated by C. A. Browne, from a late Greek alchemical poem:

> A dragon springs therefrom which, when exposed
> In horse's excrement for twenty days,
> Devours his tail till naught thereof remains.
> This dragon, whom they Ouroboros ['Tail-biter'] call,
> Is white in looks and spotted in his skin.
> And has a form and shape most strange to see.
> When he was born he sprang from out the warm
> And humid substance of united things.
> The close embrace of male and female kind
> – A union which occurred within the sea –
> Brought forth this dragon, as already said;
> A monster scorching all the earth with fire,
> With all his might and panoply displayed,
> He swims and comes unto a place within
> The currents of the Nile; his gleaming skin
> And all the bands which girdle him around
> Are bright as gold and shine with points of light.
> This dragon seize and slay with skilful art

Within the sea, and wield with speed thy knife
With double edges hot and moist, and then
His carcass having cleft in twain, lift out
The gall and bear away its blackened form,
All heavy with the weight of earthy bile;
Great clouds of steaming mist ascend therefrom
And these become on rising dense enough
To bear away the dragon from the sea
And lift him upward to a station warm,
The moisture of the air his lightened shape
And form sustaining; be most careful then
All burning of his substance to avoid
And change its nature to a stream divine
With quenching draughts; then pour the mercury
Into a gaping urn, and when its stream
Of sacred fluid stops to flow, then wash
Away with care the blackened dross of earth.
Thus having brightened what the darkness hid
Within the dragon's entrails thou wilt bring
A mystery unspeakable to light;
For it will shine exceeding bright and clear,
And, being tinged a perfect white throughout,
Will be revealed with wondrous brilliancy,
Its blackness having all been changed to white;
For when the cloud-sent water flows thereon
It cleanses every dark and earthy stain.
Thus he doth easily release himself
By drinking nectar, though completely dead;
He poureth out to mortals all his wealth
And by his help the Earth-born are sustained
Abundantly in life, when they have found
The wondrous mystery, which, being fixed
Will turn to silver, dazzling bright in kind,
A metal having naught of earthy taint,
So brilliant, clear, and wonderfully white.

This is not by any means one of the most recondite of symbolic instructions and can indeed be interpreted fairly easily. The dragon was an alloy of copper and silver made by warming the two metals with mercury ('the sea') in a vessel of fermenting horse-dung – a favourite alchemical source of gentle heat. At the end of twenty days no traces of the silver and copper remained visible, so that the dragon had bitten its tail till nothing was left. The speckled white copper-silver amalgam was then heated in an Egyptian retort ('comes into a place within the currents of the Nile') until all the mercury had distilled away and was condensed 'into a stream divine'. The liquid mercury was then poured into a receiver or 'gaping urn', and the hot liquid alloy was left in the retort, covered with a black dross of oxide. This 'earthly bile' was then removed and the bright flash of the melted metal underneath could be seen. Finally the cooled and solidified alloy was washed in running water, leaving what the alchemist thought was silver, 'dazzling bright in kind and having naught of earthly taint'. The weight of the silver-copper alloy, if the work were done carefully, need not have been much short of the combined weights of the silver and copper used, and the alchemist obviously thought that the copper had been transmuted into silver; it should, however, be noted that the alloy would not have deceived a silversmith or metallurgist of the time.

While such symbolic recipes as this are often comprehensible up to a point – though usually valueless when deciphered – there are others where the symbolism is so vague that it cannot have any precise meaning and must be regarded as a poetic alchemical effusion. The following extract is typical of this class:

> The third daye again to life he shall uprise,
> And devour byrds and beastes of the wildernesse,
> Crowes, popingayes, pyes, pecocks, and mevies [seagulls];
> The phenix, the eagle whyte, the gryffon of fearfulnesse,
> The greene lyon and the red dragon he shall distresse;
> The whyte dragon also, the antelope, unicorne, panthere,
> With other byrds, and beastes, both more and lesse,
> The basiliske also, which almost each one doth feare.

It is possible to see in this fifteenth-century passage some traces of Christian imagery, but little if any link with the laboratory of a working alchemist. In later centuries a bifurcation gradually became perceptible and grew more marked; while those whose primary aim was material transmutation still spoke much in allegory, there were alchemists or thinkers with alchemical leanings who rarely if ever lit an athanor or wielded a pestle. Thomas Vaughan (1626–66) may be included among the latter, for though in early years he had dabbled about in the laboratory he felt an aversion from what he described as 'the torture of metals'; yet that did not prevent him from employing alchemical ideas and symbolic expressions in his poems.

In pictorial allegory and symbolism the literature of alchemy is very rich; there are, in fact, treatises on the subject that consist of little else but pictures. The book of the Rabbi Abraham used by Nicholas Flamel (p. 239) was essentially graphic, as was also *Atalanta Fugiens* or 'Atalanta Fleeing' of the alchemist Michael Maier (1568–1622), an early Rosicrucian. Perhaps the most celebrated of such books is that entitled *Splendor Solis*, or 'Splendour of the Sun', of which there is a beautiful manuscript on vellum in the British Museum. This work is ascribed to one Salomon Trismosin, who lived in the sixteenth century and had adventures comparable with those of Denis Zachaire (p. 249). Some of the illustrations of the *Splendor Solis* are reproduced in plates 30, 31, 32, where much of the usual alchemical symbolism will be seen: the philosophers' egg with cocks and dragons, the appearance of the peacock's-tail colours, and the stage of the white elixir. Another work often well illustrated is entitled 'The Book of the Holy Trinity', written by an anonymous German alchemist at the time of the Council of Constance (1414–18); D. I. Duveen has described an early seventeenth-century manuscript of this treatise in an article in 'Ambix'. Plates 33, 34, 35 are reproduced from another manuscript of the work in the State Library, Munich. Plate 33 consists of five separate miniatures representing alchemical operations. First, there is depicted a leper hanged on a golden gibbet: this is the operation of calcination. Next, a leper with his hands tied behind his back is about to be decapitated by an executioner, also leprous: this is distillation. The leper attached to a gilded wheel represents coagulation, and the silver chalice and three dice, solution. The fifth miniature, of a half-woman half-serpent having a leprous bust and transfixing a leper with a golden lance, while a leprous woman stands beneath the lance, represents the extraction of philosophers' mercury from the prime matter by means of the philosophic fire. The whole plate, says Duveen, represents the exaltation of the common base metals, which are throughout considered allegorically as in a state of sin. Plate 34 represents the Grand Hermetic Androgyne trampling underfoot the four elements of the prime matter. The androgyne indicates that two opposite natures have been indissolubly united. The right-hand, masculine side of the figure is clad in blue armour heightened with silver (Jupiter – Moon); the left-hand, feminine side wears a brown robe heightened with gold (Saturn – Sun). The right-hand wing is green, heightened with gold (Venus – Sun), and the left-hand one blue, heightened with silver (Jupiter – Moon). The figure thus indicates complete intermingling and exchange of characters and properties of the two opposing principles. The figure stands in a triumphant attitude, holding in its right hand a sword (fire) and a red (Mars) crown; the left hand holds the golden crown of the Great Work.

Plate 35 shows the vessels to be used in the Great Work and the symbols of the four evangelists. The first vessel symbolizes the philosophic egg in which the material is to be placed. The second is a simple pelican used for cohobation (p. 52), and the third a double pelican for circulation; the fourth is a still with receiver. The symbols of the four evangelists represent at once the four elements, the four stages in the Work, and the four cardinal virtues.

Delightful as these and many other allegorical alchemical pictures undoubtedly are, their interpretation is often dubious. The language of symbolism affords much scope for the exercise of imagination and holds many pitfalls. Appealing rather to emotions and aspirations than to the intelligence, graphic alchemy is almost entirely esoteric.

Alchemical symbolism has recently been the subject of a profound psychological study by C. G. Jung, in his book *Psychology and Alchemy*. Jung's views are difficult to summarize, but he says that alchemy was essentially 'chemical research work into which there entered, by way of projection, an admixture of unconscious psychic material'. Our modern consciousness has had a long and gradual development, the principal factors concerned being the Church on the one hand and science on the other. Much early science, comprising remnants of the classical spirit and the classical feeling for nature, could not be accepted by the Church but found asylum in medieval natural philosophy. Religious dogma and ritual tended to draw consciousness away from its true roots in the unconscious, and alchemy, like astrology, was an attempt to maintain a bridge between the two. The symbolism of alchemy expresses the problem of the evolution of personality, and the alchemists ran counter to the Church in preferring to seek through knowledge rather than to find through faith. The Church, at least in its dogma, made the conflict between good and evil an absolute one, but there is no such sharp distinction in the psychic archetype. Hence there was an urge to effect a union of the opposites, as typified in the frequently encountered 'chemical marriage', where opposing principles are fused into a purified and incorruptible whole. The symbolic equation of Christ with the philosophers' stone may be explained as a projection of the redeemer-image, but with the reservation that the Christian earns the fruits of grace from a work already performed, while the alchemist labours 'in the cause of the divine world-soul slumbering and awaiting redemption in matter'.

It must be left to psychologists to pronounce judgement on these suggestions, but some modifications would seem to be necessary if Muslim alchemical symbolism is to be interpreted on similar lines. In any case, the practical alchemists, whether Christian or Muslim, were aiming at practical ends; and though some of them were inspired by mercenary motives, others were imbued with the true scientific spirit of inquiry: *felix qui potuit rerum cognoscere causas.*

8

PARACELSUS

ONE of the most picturesque figures in the history of Renaissance alchemy is Paracelsus, whose true name was Theophrastus Bombastus von Hohenheim. The 'Paracelsus' (together with Philippus Aureolus) was self-bestowed, to indicate that in his own opinion he was greater than Celsus, the celebrated Roman writer on medicine who lived in the first half of the first century A.D., and whose reputation as an authority on the subject ranks with those of Hippocrates and Galen. The grandiloquence is typical: Paracelsus had virtues but modesty was not one of them, and he would have had no need for a publicity agent.

He was born on 17 December 1493 at Maria-Einsiedeln near Zürich, his father Wilhelm von Hohenheim being a licentiate of medicine and the local physician. He seems to have been an only child, his mother dying at his birth or soon afterwards. Nine years later his father moved to Villach, near Klagenfurt, and practised there until his death in 1534. We know little of Paracelsus's early life, though he says that he was taught medicine and alchemy by his father, but there is little doubt that in 1514 he went to the mines and metallurgical workshops of Sigismund Fugger, in the Tyrol. Here he worked for a year and was able to assimilate much technical information concerning the precious metals and also to broaden his knowledge of alchemy, for Fugger was an enthusiastic alchemist. There is a tradition, quite likely to be well founded, that before he went to the Tyrol he had studied at the university of Basel, but another story, that he also studied at Würzburg under Hans von Trittenheim, generally known as Trithemius, is probably untrue. In the preface to one of his books, Paracelsus says that his father never ceased to help him, and that afterwards he learnt from many others, including abbots, bishops, doctors, and others, 'and I have also had great experience for a long time with many alchemists who have investigated these arts'.

Of a restless disposition, Paracelsus was constitutionally incapable of remaining long in one place. After learning all that Fugger could teach him, he set off on a rambling journey through Germany, France, Belgium, England, Scandinavia, and Italy, and may even, according to some, have visited Russia and the East. It is possible that during much of this time he was acting as an army surgeon in various campaigns, including the Venetian wars of 1521–5. He remained in Italy long enough to gain the degree of M.D. at the University of Ferrara.

While on his travels, Paracelsus associated with people of all classes and callings – physicians, alchemists, astrologers, apothecaries, miners, gypsies, and adepts of various occult arts – returning to Germany with a stock of curious miscellaneous knowledge such as few men can ever have possessed. In 1526 he took out rights of citizenship at Strasbourg and settled down in that city to practise medicine, but he had not been there long when chance presented him with a brilliant opportunity of advancement. It so happened that at that time Johann Froben or Frobenius, a prosperous printer and publisher of Basel, fell ill with a malady that baffled the local practitioners. Having heard of the new physician at Strasbourg, Froben sent for him, and Paracelsus effected a rapid and complete cure. His stock therefore stood high, and from physician he was promoted to friend. Now Erasmus, the Dutch scholar regarded as the leader of Renaissance learning in northern Europe, was at that time staying in Froben's house, which was visited also by many other scholars, including Johann Heussgen or Oecolampadius, professor of theology in Basel University. These men were much impressed by the personality and medical skill of the new physician, and recommended him to the Basel city authorities for the then vacant position of City Physician and Professor of Medicine, an offer that was at once accepted. Not long afterwards, Erasmus himself fell ill and wrote to Paracelsus as follows: 'I cannot offer thee a reward equal to thy art and knowledge – I surely offer thee a grateful soul. Thou hast recalled from the shades Frobenius who is my other half: if thou restorest me also thou restorest each through the other. May fortune favour that thou remain in Basel.'

In point of fact Paracelsus held the appointment for only two years. His originality led him to decry the virtues of the old-fashioned herbal remedies then universally employed, and to extol the powers of medicines made from minerals; moreover he offended conservative opinion by lecturing in German instead of in Latin. Matters were exacerbated by the abuse that Paracelsus, and the followers that he soon attracted, hurled at those who thought differently.

Paracelusus [says Robert Burton (1577–1640) in his *Anatomy of Melancholy*], and his Chemistical followers, as so many Prometheuses, will fetch fire from Heaven, will cure all manner of diseases with minerals, accounting them the only Physick ... Paracelsus calls Galen, Hippocrates, and all their adherents, infants, idiots, Sophisters, &c. Away (he says) with those who jeer at Vulcanian metamorphoses, ignorant sprouts, backward and stubborn nurslings, &c., not worthy the name of Physicians, for want of these remedies; and brags that by them he can make a man live 160 years, or to the world's end; with their Alexipharmacums, Panaceas, Mummias, Weapon Salve, and such Magnetical cures, Lamps of Life and Death, Balsams, Baths of Diana, Magico-Physical Electrum, Martian Amulets, &c. What will not he and his followers effect? He brags moreover that he was the First of Physicians, and did more famous cures than all the Physicians in Europe besides; a drop of his preparations should go further than a dram or ounce of theirs, those loathsome and fulsome filthy potions, heteroclitical pills (so he calls them), horse medicines, at the sight of which the Cyclops Polyphemus would shudder.

A man more unsuited to hold public office than 'marvellous Paracelsus, always drunk and always lucid, like the heroes of Rabelais', can hardly be imagined. He signalized his appointment as City Physician by publicly burning the works of Avicenna and Galen in a brass pan with sulphur and nitre, to show his contempt of orthodox medicine and to emphasize the fact that his doctrines were his own and wholly original.

If your physicians [he said] only knew that their prince Galen – they call none like him – was sticking in Hell, from whence he has sent letters to me, they would make the sign of the cross upon themselves with a fox's tail. In the same way your Avicenna sits in the vestibule of the infernal portal; and I have disputed with him about his *aurum potabile*, his Tincture of the Philosophers, his Quintessence, and Philosopher's stone, his Mithridatic, his Theriac, and all the rest. O you hypocrites, who despise the truths taught you by a great physician, who is himself instructed by Nature, and is a son of God himself! Come then, and listen, impostors who prevail only by the authority of your high positions! After my death, my disciples will burst forth and drag you to the light, and shall expose your dirty drugs, wherewith up to this time you have compassed the death of princes, and the most invincible magnates of the Christian world. Woe for your necks in the day of judgement! I know that the monarchy will be mine. Mine, too, will be the honour and glory. Not that I praise myself: Nature praises me. Of her I am born; her I follow. She knows me, and I know her. The light which is in her I have beheld in her; outside, too, I have proved the same in the figure of the microcosm, and found it in that universe.

As may easily be imagined, such conduct did not increase the popularity of this theatrical medical officer of health. But his vituperation did not confine itself to general attacks on the whole body of physicians: individual members as well felt the venom of his tongue. To one who had ventured to disagree with him he replied in the following terms:

So then, you wormy and lousy Sophist, since you deem the Monarch of Arcana a mere ignorant, fatuous, and prodigal quack, now, in this mid age, I determine in my present treatise to disclose the honourable course of procedure in these matters. ... for the use and honour of all who love the truth, and in order that all who despise the true arts may be reduced to poverty.

Before long Paracelsus became an object of hatred to all the druggists and apothecaries in the city, as well as to his brother physicians. Attempts were made to get rid of him and his lectures were ruined by noisy interruptions, but to their credit the city authorities supported him and would not yield to the clamour. However, matters were at length brought to a crisis. A prominent citizen of Basel, Canon Lichtenfels, fell ill and offered a fee of 100 gulden to any physician who could cure him. Paracelsus accepted the offer and cured his patient, who then refused to pay the agreed fee. Paracelsus was not the man to submit tamely to such treatment and took the matter to court. For some reason or other – possibly a legal quibble, since he seems to have had right on his side – judgment was given against him, an outcome at which he was so infuriated as to surpass his own virtuosity in execration of the startled magistrates. Warned that this flow of scurrilous and slanderous epithets had rendered him liable to severe punishment for contempt of court, he left Basel secretly and hurriedly, setting out once more upon a life of wandering through Germany and Austria. We hear of him at Colmar, Esslingen, Nuremberg, and St Gallen; an inhabitant of the last-named town described him as 'ceaselessly, ceaselessly writing'. During this time his fortunes had sunk very low, so much so that his unkempt and tramp-like appearance led to his being refused admission to the town of Innsbruck. As he himself said, 'The burgomaster of Innsbruck has probably seen doctors in silks at the courts of princes but not sweltering in rags in the sun.' At last he was invited to Salzburg by the Prince Palatine, Archbishop Duke Ernst of Bavaria, himself a student of the occult arts. Here he seems to have found a peaceful and congenial atmosphere, but he was not long to enjoy it. Arriving at Salzburg in April 1541, he died there on 24 September of the same year at the age of forty-eight; a comparatively young man, he was physically worn out by his restless and strenuous life. He was buried in the churchyard of St Sebastian, and his epitaph read:

Here lies buried Philippus Theophrastus, distinguished Doctor of Medicine, who with wonderful art cured dire wounds, leprosy, gout, dropsy and other contagious diseases of the body, and who gave to the poor the goods which he obtained and accumulated. In the year of our Lord 1541, the 24th September, he exchanged life for death.

Paracelsus's enemies circulated the story that his death had been due to a drunken orgy, a story that received some apparent confirmation when on exhumation of his body at the beginning of the nineteenth century his skull was found to show a fracture that might have been caused by a blow or a fall. Further examination in the 1880s, however, revealed that the fissure was a result of rickets, and the fact that Paracelsus made his will (still preserved) only three days before he died indicates that he knew his end to be approaching.

The first collected edition of Paracelsus's works appeared in ten volumes in 1589–91, carefully compiled and edited by Johannes Huser; it was published at Basel. Second and third editions of this collection were published at Strasbourg in 1603 and 1616, and it was translated into modern German by Aschner in 1926–30. For the purpose of his collection Huser travelled extensively in Germany and Austria in search of early editions and manuscripts, and was methodical enough to prefix a note to each work in his edition saying when it is printed from a manuscript in Paracelsus's own hand and when from an authentic copy. He also includes some works which he thinks are probably not genuine, and in this class he places all those that deal exclusively with alchemy. Huser's opinion that Paracelsus did not write any books specifically upon alchemy is shared by most later scholars; hence his views on the subject have to be extracted from his voluminous medical and philosophical works, which do indeed contain a great deal of alchemy. The difficulty is enhanced by Paracelsus's habit of writing in a mixture of German, dog Latin, and words of his own invention (p. 172), but it is now possible to form an idea of his main lines of thought.

In the first place, Paracelsus, though not denying the possibility of transmutation, regarded this aspect of alchemy as of secondary importance. 'Many have said of alchemy,' he writes, 'that it is for making gold and silver. For me such is not the aim, but to consider only what virtue and power may lie in medicines.' His practical object was to use alchemical processes for the preparation of therapeutic substances, principally from inorganic sources, and he thus inaugurated the art or science of iatrochemistry (medical chemistry) or, as we should now say, chemotherapy.

The theoretical or philosophical background to this object has as its fundamental principle the unity of all things, explained by a recent student of Paracelsan literature, T. P. Sherlock, as follows:

There is a unity of heaven and earth, which is thought of partly in terms of the real or supposed influence which the sun, the stars, and the planets exert upon the life of plants, animals, and men and on medicines derived from these, and partly in terms of the repetition of pattern which gives rise to the notions of macrocosm and microcosm. The same influence makes him view Nature, in her operations that bring about life, death, sickness, and health, as a kind of 'world-alchemist'; man in his function of preparer of remedies becomes the alchemist of nature, while within man dwells a further 'archeus' or alchemist who works in the stomach, which is, as it were, his laboratory.

Paracelsus believed that sickness and health are controlled by astral influences, and that sickness can be driven out and health restored by 'arcana' or secret remedies. The function of the arcana is to restore a celestial harmony between an inner 'astrum' or star in man and a heavenly astrum; they must therefore 'reach out to heaven', that is, be volatile and incorporeal. The actual medicine must of course be material, but the arcanum it contains is spiritual. Physicians must know about the stars in order to know the causes of sickness, and about alchemy in order to know how to prepare the arcana.

These ideas are expressed in Paracelsus's *Panagranum*, from which the following passage is taken, in Sherlock's translation:

It is not the physician that controls and directs, but heaven, by the stars; and therefore the medicine must be brought into an airy form in such a manner that it may be directed by the stars. For what stone is lifted up by the stars? None save the volatile. Hence many have looked for the *quintum esse* in alchemy which is nothing else but that thus the four elements are taken away from the arcanum, and what remains afterwards is the arcanum. This arcanum furthermore is a chaos [gas], and it is possible to carry it to the stars like a feather before the wind. Now the preparation of the medicine should be done in this way. The four elements are taken away from the arcanum, and then it should be known what astrum is in this particular arcanum, and what is the astrum of this particular disease, and what astrum is in the medicine against the disease. This is where the directing comes in. If you give a medicine, the stomach has to prepare it, and it is the alchemist. If it is possible for the stomach so to manage it that the astra accept each other, the medicine is directed;

if not it remains in the stomach and goes out through the stool. For what is more noble in a doctor than a knowledge of the concordance of both astra? – for there lies the basis of all diseases. Now alchemy is the outer stomach which prepares for the stars its own. The purpose of alchemy is not, as it is said, to make gold and silver, but in this instance to make arcana and direct them against diseases; as this is the outcome, so it is also the basis. For all these things conform to the instruction and test of nature. Hence nature and man, in health and sickness, need to be joined together, and to be brought into mutual agreement. This is the way to heal and restore to health. All this shall be achieved by alchemy, without which these things cannot be done.

Paracelsus attached much importance to the method of preparation of remedies. He believed that a particular substance may contain many different potential powers, and that nature orders alchemy to prepare it in one way for one disease and in another way for a different disease.

The doctor then sheds light on the matter, for he knows the cause, and with it the manner of cooking and preparing. But what light do you shed, you doctors of Montpellier, Vienna, and Leipzig? About as much light as a Spanish fly in a dysentery stool!

In general terms, Paracelsus understands by alchemy any process that transforms raw material into a finished product. God has made all things, and these have a natural end to which they will come by the power of nature within them; but God has also assigned to man the power of transforming things from this natural or raw state to a condition fit for man's use. Thus God has given iron ore, but it must be worked up by the smelter and the smith, who are therefore alchemists. The processes of milling and baking are examples of 'outer' alchemy; the conversion of flour in the alimentary canal is an 'inner' alchemy. Paracelsus thus uses the word alchemy in a different sense from that understood by his predecessors and contemporaries – a typical habit of his, and one that he supplements by inventing new words when the whim takes him. Thus he quite arbitrarily transferred the name alcohol from black eye-paint (*al-kohl*) to spirit of wine, and there it has remained ever since. He also invented the word zinc, another innovation that persisted, and the name alkahest for a supposed universal solvent. For his own style of alchemy, medical or iatric, he coined the name spagyric, which was afterwards sometimes used for alchemy in general.

Though Paracelsus went to extreme lengths in abuse of his opponents, we should remember that the apothecaries of that time usually had no knowledge of chemistry, preparing their medicines from roots, leaves, fruits, syrups, and the like after the fashion of a village housewife. The physicians were in no better case. 'They think it suffices', says Paracelsus, 'if, like apothecaries, they jumble a lot of things together and say "*Fiat unguentum*".... Yet if medicine were handled by artists [chemists], a far more healthy system would be set on foot.' For the few apothecaries and physicians who were enlightened enough to study chemistry and brave enough to apply chemical remedies, he had the warmest praise:

They do not consort with loafers or go about gorgeous in satins, silks, and velvets, gold rings on their fingers, silver daggers hanging at their sides, and white gloves on their hands, but they tend their work at the fire patiently day and night. They do not go promenading, but seek their recreation in the laboratory, wear plain leathern dress and aprons of hide upon which to wipe their hands, thrust their fingers amongst the coals, into dirt and rubbish and not into golden rings. They are sooty and dirty like the smiths and charcoal-burners, and hence make little show, make not many words nor gossip with their patients, and do not highly praise their own remedies, for they well know that the work must praise the master, not the master his work. They well know that words and chatter do not help the sick nor cure them ... Therefore they let such things alone and busy themselves with working with their fires and learning the steps of alchemy. These are distillation, solution, putrefaction, extraction, calcination, reverberation, sublimation, fixation, separation, reduction, coagulation, tinction, and the like.

Such plain speaking, with its accompanying good sense, gradually had the effect of making chemistry, for the future, an indispensable part of a medical training. Physicians began to break free from tradition, and chemistry found a place in the curricula of colleges and medical schools. Paracelsus had re-orientated the old alchemy, and his call to the alchemists to prepare new remedies soon had results.

On what we may call orthodox alchemy, Paracelsus has comparatively little to say. Believing that the universe as a whole and all the objects in it were endowed with life, he peopled the intermediate state between the material and the immaterial with beings consisting of a body and spirit but no soul; such were the sylphs (another word of Paracelsan invention) of the air, the nymphs of water, and the salamanders of fire. As to material substances, he considered them to be ultimately composed of the four Aristotelian elements, but immediately of three primary bodies, *tria prima*, namely salt (body), sulphur (soul), and mercury (spirit). He was thus taking over a previously existing modification of the old sulphur-mercury theory of metals, extended so as to apply to all substances, metallic and non-metallic, animal, and vegetable. Salt was the principle of incombustibility and non-volatility; mercury was the principle of fusibility and volatility; and sulphur was the principle in virtue of which substances are inflammable. The theory was not intended to be taken literally; the 'sulphur' in wood, for example, is not the same as the 'sulphur' in lead, and neither of them is to be conceived as very closely resembling ordinary sulphur. The *tria prima*, or, as they are often known, 'hypostatical principles', are indeed nothing more than abstractions of qualities, and therefore differ essentially in character from the elements of modern chemistry. Paracelsus himself says:

You should know that all seven metals originate from three materials, namely from mercury, sulphur, and salt, though with different colours. Therefore Hermes has said not incorrectly that all seven metals are born and composed from three substances, similarly also the tinctures and philosophers' stone. He calls these three substances spirit, soul, and body. But he has not indicated how this is to be understood nor what he means by it. Although he may perhaps have known, yet he has not thought to say it. I therefore do not say that he has erred, but only kept silent. But that it be rightly understood what the three different substances are that he calls spirit, soul, and body, you should know that they mean not other than the three *principia*, that is, mercury, sulphur, and salt, out of which all seven metals originate. Mercury is the spirit, sulphur is the soul, salt the body.

But as there are many kinds of fruit, so there are many kinds of sulphur, salt, and mercury. A different sulphur is in gold, another in silver, another in lead, another in iron, tin, &c. Also a different one in sapphire, another in the emerald, another in the ruby, chrysolite, amethyst, magnets, &c. Also another in stones, flint, salts, spring-waters, &c. And not only so many kinds of sulphur but also so many kinds of salt, different ones in metals, different ones in gems, stones, others in salts, in vitriols, in alum. Similarly with mercuries, a different one in the metals, another in gems, and as often as there is a species there is a different mercury. Of one nature is sulphur, of one nature salt, of one nature mercury. And further they are still more divided, as there is not merely one kind of gold but many kinds of gold, just as there is not merely one kind of pear or apple but many kinds. Therefore there are just as many different kinds of sulphurs of gold, salts of gold, and mercuries of gold.

In the human body, the three hypostatical principles produced disease if they were not properly balanced; thus excess of sulphur caused fever and plague while too little gave rise to gout. Excess of mercury was the cause of paralysis and depression, and excess of salt produced diarrhoea and dropsy. Even when the total balance in the body was correct, illness might result from excess or deficiency in particular organs; thus an internal transference of sulphur might give rise to delirium.

There are many passages throughout Paracelsus's books that indicate his practical familiarity with chemical operations, but they indicate equally clearly that he had only the vaguest ideas as to what went on in the operations. He often seems to think that a particular virtue or property could be induced in whatever substance if the appropriate procedure were followed; thus a sequence of calcination, solution, distillation, coagulation, and so forth might produce a very different end-product, if the original material were copper, from that obtained using iron as the original material – but no matter: the two end-products, however unlikely they appeared, would be endowed with the same virtue because they were prepared in the same way. Yet, at other times Paracelsus attempts to classify substances as well as operations, and though it is improbable that the idea of chemical purity ever occurred to him, his views on the 'arcana' did unwittingly lead in that direction. He believed that the freer an arcanum was from contaminates, the more powerful its action would be, and his followers were thus led to invent methods of purification and tests for purity. Such tests were at first qualitative, but numerous unfortunate incidents consequent upon the introduction into medicine of such poisons as arsenic, antimony, mercury, and their compounds brought the realization that purity must be measured quantitatively for clinical purposes. Thus Paracelsus and his disciples not only provided alchemy with a new outlook and a new purpose, but unintentionally paved the way to a conception of a basic law of chemistry, namely that all specimens of a given chemical individual have identical compositions.

Paracelsus's alchemical views and speculations led him into the fields of psychology and psychiatry, where he showed much originality; but to follow him here would mean a lengthy digression. According to Jung, Paracelsan alchemy in its psychological aspect is a path of salvation for the soul – the process of becoming the perfect man – and Paracelsus believed that when the conscious will and intellect were flooded with the super-personal *lumen naturae*, or light of Nature, life's destiny was fulfilled.

9

SOME ENGLISH ALCHEMISTS

CHAUCER AND ALCHEMY

HOW widespread the practice of alchemy had become in England by the second half of the fourteenth century may be gauged by the account Chaucer (1340?–1400) gives of it in his *Canterbury Tales*. 'The Canon's Yeoman's Tale' is the story of an apparently honest but deluded alchemical canon, and is related by his yeoman-acting-laboratory-assistant. From the descriptions given of alchemical materials and processes it is evident that Chaucer had studied the art with close attention, and certain rather bitter remarks suggest that he had himself lost time and money in unsuccessful attempts at transmutation.

The Canon and his yeoman overtake the pilgrims as they are riding along the road to Canterbury, and the yeoman explains that he had seen them leave their hostelry that morning and had told his master. The Canon said he would like to join the company, so he and the yeoman set out as soon as they were ready and rode at a good pace to overtake it. When they arrived, the pilgrims were curious to have some information about the newcomers, and the host inquired of the yeoman about his master, 'Is he a clerk or noon? tel what he is.'

> 'Nay, he is gretter than a clerk, y-wis,'
> Seyde this yeman, 'and in wordes fewe,
> Host, of his craft som-what I wol yow shewe.
> I seye, my lord can switch subtilitee. . . .
> That al this ground on which we been ryding,
> Til that we come to Caunterbury toun,
> He coude al clene turne it up-so-doun,
> And pave it al of silver and of gold.'

Such an opening remark made the host eager for more, but he first expressed surprise that the Canon, possessing such a marvellous power, should be dressed so shabbily – his 'oversloppe' was not worth a mite and was 'al baudy and to-tore':

> Why is thy lord so sluttish, I thee preye,
> And is of power better cloth to beye,
> If that his dede accorde with thy speche?
> Telle me that, and that I thee biseche.

The yeoman explained that the Canon was not only learned but too learned, and, as often happened, misused his learning. The host became 'curiouser and curiouser', especially when the yeoman added that they lived secretively in an unsavoury part of the suburbs of a town, and was moved to ask the yeoman why his face was so discoloured:

> 'Peter!' quod he, 'god yeve it harde grace,
> I am so used in the fyr to blowe,
> That it hath chaunged my colour, I trowe.
> I am nat wont in no mirour to prye,
> But swinke [toil] sore and lerne multiplye [p. 152].
> We blondren ever and pouren in the fyr,
> And for al that we fayle of our desyr,
> For ever we lakken our conclusion.'

At this moment the Canon observed that they were talking about him, and ordered his servant to desist:

> 'Hold thou thy pees, and spek no wordes mo,
> For if thou do, thou shalt it dere abye;
> Thou sclaundrest me heer in this companye,
> And eek discoverest that thout sholdest hyde.'

But the host urged him to continue and reck not a mite for the Canon's threats, and, as the yeoman did not require much persuasion, the Canon 'fledde awey for verray sorwe and shame'. With that, the yeoman settled down to his story:

> With this chanoun I dwelt have seven yeer,
> And of his science am I never the neer.
> Al that I hadde, I have y-lost ther-by;
> And god wot, so hath many mo than I.
> Ther I was wont to be right fresh and gay
> Of clothing and of other good array,
> Now may I were an hose upon myn heed;

And wher my colour was bothe fresh and reed,
Now is it wan and of a leden hewe;
Who-so it useth, sore shal he rewe.
And of my swink yet blered is myn yë,
Lo! which avantage is to multiplye!
That slyding science hath me maad so bare,
That I have no good, wher that ever I fare;
And yet I am endetted so ther-by
Of gold that I have borwed, trewely,
That whyl I live, I shal it quyte never.
Let every man be war by me for ever!
What maner man that casteth him ther-to,
If he continue, I holde his thrift y-do.
So helpe me god, ther-by shal he nat winne,
But empte his purs, and makes his wittes thinne.

The yeoman spoke feelingly, having at first had sufficient faith in his master to risk his own capital, and to work for seven years in the laboratory in the hope of future success but to the detriment of his health. His hearers were now thoroughly absorbed in his story, so he went on to describe some of the secrets of the smoky operations carried out in the dingy and untidy back room. First he recites a catalogue of apparatus and materials:

As bole armoniak, verdegree, boras,
And sondry vessels maad of erthe and glas,
Our urinales and our descensories,
Violes, croslets [crucibles], and sublymatories,
Cucurbites, and alembykes eek,

. . . . .

Watres rubifying and boles galle,
Arsenik, sal armoniak, and brimstoon;
And herbes coude I telle eek many oon,
As egremoine, valerian, and lunaire,
And othere swiche, if that me liste tarie.
Our lampes brenning bothe night and day,
To bringe aboute our craft, if that we may.
Our fourneys [furnaces] eek of calcinacioun,
And of watres albificacioun.

. . . . .

And divers fyres maad of wode and cole [charcoal];
Saltarte, alkaly, and sal preparat,
And combust materes and coagulat,
Cley maad with hors or mannes here, and oile
Of tartre, alum, glas, berm, woret, and agoile,
Resalgar, and our materes enbibing.

No doubt the pilgrims were somewhat bewildered by these 'clergial' and 'queynte' terms, but the yeoman was enjoying himself as the centre of attention and proceeded to give further enlightenment:

I wol yow telle, as was me taught also,
The foure spirites and the bodies sevene,
By ordre, as ofte I herde my lord hem nevene [name].
The firste spirit quik-silver called is,
The second orpiment, the thridde, y-wis,
Sal armoniak, and the ferthe brimstoon.
The bodies sevene eek, lo! hem heer anoon:
Sol gold is, and Luna silver we thrape,
Mars yren, Mercurie quik-silver we clepe,
Saturnus leed, and Jupiter is tin,
And Venus coper, by my fader kin!

The yeoman then expatiates on the foolishness of alchemists in general, saying that although the search is hopeless they will not desist in their efforts to discover the philosophers' stone. They spend their last penny in buying drugs and minerals, and care not about threadbare clothes or the insalubrious atmosphere of their lair, and

Evermore, wher that ever they goon,
Men may hem knowe by smel of brimstoon;
For al the world, they stinken as a goot;
Her savour is so rammish and so hoot,
That, though a man from hem a myle be,
The savour wol infecte him, trusteth me.

But let us pass that, he says, and get on with the tale:

Er than the pot be on the fyr y-do,
Of metals with a certein quantitee,
My lord hem tempreth, and no man but he –
Now he is goon, I dar seyn boldely –
For, as men seyn, he can don craftily [cleverly];
Algate [at least] I woot wel he hath swich a name,
And yet ful ofte he renneth in a blame;
And wite ye how? ful ofte it happeth so,
The pot to-breketh, and farewel! al is go!

After one such accident, fellow alchemists were ready to offer explanations of why the pot broke. One said it had been left on the fire too long; another, that the bellows had not been properly worked ('than was I fered', says the yeoman, 'for that was myn office'); a third, that the mixture had not been tempered aright; and a fourth, that the fire was not made of beech-charcoal, as it should have been. The Canon grew impatient with these useless arguments, and said:

... ther is na-more to done,
Of thise perils I wol be war eft-sone;
I am right siker [certain] that the pot was crased [cracked]
Be as be may, be ye no-thing amased;
As usage is, lat swepe the floor as swythe [at once],
Plukke up your hertes, and beth gladde and blythe.

The sweepings were sifted on to a piece of canvas, and:

'Pardee,' quod oon, 'somwhat of our metal
Yet is ther heer, though that we han nat al.
Al-though this thing mishapped have as now,
Another tyme it may be wel y-now.'

The yeoman gloomily ends the first part of his story by saying that 'we concluden evermore amis', and that although when the alchemists are in debate every man thinks he is as wise as Solomon, when it comes to the proof they are all either fools or rogues. His master the Canon is in the first of these categories, but the yeoman now tells of another canon, whose infinite deceits no man could write, even if he lived to be as old as Methuselah. One day this canon called on a good-natured priest and begged the loan of a small sum of money:

'Lene me a mark,' quod he, 'but dayes three,
And at my day I wol it quyten thee.
And if so be that thou me finde fals,
Another day do hange me by the hals [neck]!'

The priest readily handed over the mark, and, true to his word, the canon returned it on the day agreed. The perennial confidence trick had been successfully launched. The priest was gratified at the canon's promptness in repaying the loan, and said to him:

... no-thing anoyeth me
To lene a man a noble, or two or three,
Or what thing is in my possessioun,
When he so trewe is of condicioun
That in no wyse he breke wol his day;
To swich a man I can never seye nay.

The canon protested that he was the soul of honour, invariably truthful and always punctilious in settling his debts. But as the priest had been so kind to him, he would like to show his appreciation by demonstrating a 'maistrie' or transmutation.

'Ye,' quod the preest, 'ye, sir, and wol ye so?
Marie! ther-of I pray yow hertely!'

The yeoman puts in an aside:

Noght wiste this preest with whom that he delte,
Ne of his harm cominge he no-thing felte.
O sely preest! O sely innocent!

For the canon was an out-and-out swindler, a hundred times more subtle than the yeoman's master; the yeoman can scarcely bring himself to speak of such deceits:

Of his falshede it dulleth me to ryme.
Ever whan that I speke of his falshede,
For shame of him my chekes wexen rede;
Algates, they biginnen for to glowe,
For reednesse [redness] have I noon, right wel I knowe,
In my visage; for fumes dyverse
Of metals, which ye han herd me reherce,
Consumed and wasted han my reednesse.

To return to the story: the canon asked the priest to send his servant for three ounces of mercury, which was soon procured. The servant was then instructed to fetch coal for the fire, and meanwhile the alchemist extracted a crucible from his bosom and shewed it to the priest.

> 'This instrument,' quod he, 'which that thou seest,
> Tak in thyn hand, and put thy-self ther-inne
> Of this quik-silver an ounce, and heer beginne,
> In the name of Crist, to wexe a philosofre.
> Ther been ful fewe, whiche that I wolde profre
> To shewen hem thus muche of my science.
> For ye shul seen heer, by experience,
> That this quik-silver wol I mortifye
> Right in your sighte anon, withouten lye,
> And make it as good silver and as fyn
> As ther is any in your purs or myn,
> Or elleswher, and make it malliable;
> And elles, holdeth me fals and unable
> Amonges folk for ever to appere!
> I have a poudre heer, that coste me dere,
> Shal make al good, for it is cause of al
> My conning [learning], which that I yow shewen shal.'

The alchemist then suggested that the servant should be dismissed and the door locked, so that the work could be put in hand. The priest set the crucible on the fire, and the alchemist added to it a little powder – what it was, the yeoman says he did not know, but it was 'nat worth a flye'. Next the charcoal had to be properly arranged above the crucible, and while the priest was attending to this matter, the canon slyly produced a piece of beech-charcoal,

> In which ful subtilly was maad an hole,
> And ther-in put was of silver lymaille [filings]
> An ounce, and stopped was, with-outen fayle,
> The hole with wex, to kepe the lymail in.
> And understondeth, that this false gin
> Was not maad ther, but it was maad bifore.
>
> . . . . .

But taketh heed now, sir, for goddes love!
He took his cole of which I spak above,
And in his hond he baar it prively.
And whyls the preest couchede busily
The coles, as I tolde yow er this,
This chanoun seyde, 'freend, ye doon amis;
This is nat couched as it oghte be;
But sone I shal amenden it,' quod he.
'Now lat me medle therwith but a whyle,
For of yow have I pitee, by seint Gyle!
Ye been right hoot, I see wel how ye swete,
Have heer a cloth, and wype awey the wete.'
And whyles that the preest wyped his face,
This chanoun took his cole with harde grace,
And leyde it above, up-on the middeward
Of the croslet, and blew wel afterward,
Til that the coles gonne faste breene.

The priest had missed this trickery, and anxiously awaited the result of the experiment. When the alchemist saw his time, he withdrew the crucible from the fire and plunged it into a vessel of water; then he told the priest to put his hand into the water and take out what was there. It was silver: and, as the yeoman exclaims,

What, devel of helle! sholde it elles be?
Shaving of silver silver is, pardee!

Such marvellous success excited the priest beyond measure, and he implored the canon to teach him 'this disciplyne and this crafty science'; but the canon said he would first make a second test, in order that the priest should thoroughly understand the procedure and if necessary be able to carry it out unassisted. On this occasion, the silver filings were concealed in a hollow stick closed with wax, with which the canon stirred the red-hot charcoal above the crucible at an appropriate moment:

Whan that this preest thus was bigyled ageyn,
Supposing noght but trouthe, soth to seyn,
He was so glad, that I can nat expresse
In no manere his mirthe and his gladnesse.

The canon, however, had a third card to play. He asked for some copper, and subjected it to the same treatment in the crucible. The molten metal was poured into a mould which was then set in the pan of water, but the canon had previously prepared an ingot of silver of the same size and shape as that produced by the mould, and had hidden it in his sleeve. He put his hand in the water to stir it round, and while doing so removed the copper ingot and substituted the silver one for it. Then he told the priest to fish out the ingot – and joy was unconfined.

To place matters beyond doubt, the canon, an artist in crime though 'feendly bothe in herte and thoght', suggested that they should now take the specimens of alchemical silver to a goldsmith for assay. They did so, and the ingots 'weren as hem oghte be', whereupon:

> This sotted preest, who was gladder than he?
> Was never brid gladder agayn the day,
> Ne nightingale, in the sesoun of May.

'For love of God,' he said to the canon, 'what shall this receipt cost?' The reply was that it was dear, for the canon and a certain friar were the only two people in England that knew it. 'No matter,' said the priest. 'For God's sake tell me what I must pay.' Thus pressed, the canon repeated that the cost was high, but that, because of the priest's earlier kindness to him, he should have the recipe for a mere forty pounds. The priest was only too happy to pay this sum, and handed it over immediately, after which the knave and his dupe parted with expressions of mutual good will. But:

> ... whan that this preest sholde
> Maken assay, at swich tyme as he wolde
> Of this receit, far-wel! it wolde nat be!
> Lo, thus byjaped and bigyled was he!

Chaucer ends his story with a clever little pastiche of an alchemist 'revealing the secret' to an aspirant:

> ... ther was a disciple of Plato,
> That on a tyme seyde his maister to,
> As his book Senior [p. 102] wol bere witnesse,
> And this was his demande in soothfastnesse:

'Tel me the name of the privy stoon?'
And Plato answerde unto him anoon,
'Tak the stoon that Titanos men name.'
'Which is that?' quod he. 'Magnesia is the same,'
Seyde Plato. 'Ye, sir, and is it thus?
This is *ignotium per ignotious.*
What is Magnesia, good sir, I yow preye?'
'It is a water that is maad, I seye,
Of elementes foure', quod Plato.
'Tel me the rote, good sir,' quod he tho,
'Of that water, if that it be your wille?'
'Nay, nay,' quod Plato, 'certein, that I nille [won't].'

Apart from Plato and Senior, the only authorities mentioned by Chaucer are Hermes and Arnold of Villanova (whose 'Rosary' is quoted).

As for the disillusioned yeoman, his parting advice is to let alchemy alone:

Thanne conclude I thus; sith god of hevene
Ne wol nat that the philosophres nevene
How that a man shal come un-to this stoon,
I rede, as for the beste, lete it goon.

GEORGE RIPLEY

One of the most celebrated of English alchemists was George Ripley, who was born at the village of Ripley, near Harrogate, early in the fifteenth century and died about 1490. He studied alchemy and other subjects in Rome, at Louvain, and on the island of Rhodes, where he was the guest of the Knights of St John of Jerusalem. The Knights had captured Rhodes in 1310, with the assistance of the Genoese, and held it as an outpost against the Turks. No doubt to demonstrate Ripley's proficiency in alchemy, a story was reported to the effect that he gave the Knights no less a sum than £100,000 yearly to support their warlike efforts against the infidel. By 1471 he was back in England, a canon in the Augustinian priory at Bridlington, where the fumes and unpleasant odours emanating from his alchemical laboratory proved a nuisance to the rest of the community.

Like the French adept Christopher of Paris, of whom he was a contemporary, Ripley was one of the first to popularize the alchemical teachings ascribed to Ramon Lully (p. 126), but he also wrote many original works on the subject. His 'Compound of Alchemie' (1470–1), written in English, was dedicated to Edward IV and was first printed in 1591; it became very popular and numerous manuscripts of it exist. In 1476 he dedicated his *Medulla alchimiae* ('Marrow of Alchemy') to George Nevill, archbishop of York (p. 193), and he also wrote a song or *Cantilena*, a sixteenth-century translation of which from the original Latin into English has been published by Sherwood Taylor; it purports to explain the alchemical mystery in an allegorical form but is even more obscure than the generality of such writings.

Ripley agrees with the majority of alchemists that the progress of the Work can be followed by observing the succession of colours occurring in it. This, he says, should be as follows:

Pale, and Black, wyth falce Citryne, unparfayt White & Red,
Pekoks fethers in color gay, the Raynbow whych shall overgoe
The Spottyd Panther wyth the Lyon greene, the Crowys byll
  bloe as lede;
These shall appere before the parfyt Whyte, & many other moe
Colours, and after the parfayt Whyt, Grey, and falce Citrine also:
And after all thys shall appere the blod Red invaryable,
Then hast thou a Medcyn of the thyrd order of hys owne kynde
  Multyplycable.

Colours are also mentioned in the 'Vision' that Ripley once saw, in which again alchemical secrets were made clear to him, if not to us:

When busie at my booke I was upon a certeine night,
This Vision here exprest appear'd unto my dimmed sight,
A *Toade* full rudde I saw did drinke the juice of grapes so fast,
Till over charged with the broth, his bowells all to brast;
And after that from poysoned bulke he cast his venome fell,
For greif and paine whereof his Members all began to swell,
With drops of poysoned sweat approaching thus his secret Den,
His cave with blasts of fumous ayre he all be-whyted then;

And from the which in space a golden humour did ensue,
Whose falling drops from high did staine the soile with ruddy hew:
And when this Corps the force of vitall breath began to lacke,
This dying *Toade* became forthwith like Coale for colour blacke:
Thus drowned in his proper veynes of poysoned flood,
For tearme of eightie dayes and fowre he rotting stood:
By tryall then this venome to expell I did desire,
For which I did committ his carkase to a gentle fire:
Which done, a wonder to the sight, but more to be rehear'st,
The *Toade* with Colours rare through every side was pear'st,
And White appeared when all the sundry hewes were past,
Which after being tincted Rudde, for evermore did last.
Then of the venome handled thus a medicine I did make;
Which venome kills and saveth such as venome chance to take.
Glory be to him the graunter of such secret wayes,
Dominion, and Honour, both with Worship, and with Prayes.

AMEN.

The permanence of the colours imparted to transmuted metals by the elixirs, white for silver and red for gold, is a sign that the transmutation has been completely successful. As Ripley reiterates in another place:

Which Tinctures when they by craft are made parfite,
So dieth [dyeth] Mettalls with Colours evermore permanent,
After the qualitie of the Medycine Red or White;
That never away by eny Fire, will be brente.

Ripley's reputation as an accomplished alchemist was very high, not only in his lifetime but for generations later. Even as late as 1678 there appeared a book entitled 'Ripley Reviv'd: or an exposition upon Sir George Ripley's Hermetico-Poetical Works'; this was published in London but its authorship is unknown. Ripley is also said to have been the instructor of Thomas Norton, author of 'The Ordinall of Alchimy', a work next to be described; and at least indirectly of William Holway (otherwise known as Gibbs), Prior of Bath Abbey at the time of the dissolution. Holway (p. 199) spent much money on the reconstruction of the Abbey, which had been destroyed in 1090 and was not completely restored until the early seventeenth century, so no doubt he was attracted to the study of alchemy by the hope of obtaining further funds for the rebuilding.

THE ORDINALL OF ALCHIMY

There are in the British Museum several manuscripts of an alchemical poem called 'The Ordinall of Alchimy'. This poem was first published in Latin translation, in a collection of three alchemical tracts edited by Michael Maier and printed at Frankfurt-am-Main in 1618. The original English version was not published until 1652, when Elias Ashmole (1617–92), founder of the Ashmolean Museum at Oxford, included it in his *Theatrum Chemicum Britannicum.* This work, now seldom offered for sale, comprises 'Severall Poeticall Pieces of our Famous English Philosophers, who have written the Hermetique Mysteries in their owne Ancient Language', and is a mine of information on early English alchemy and alchemists. The 'Ordinal' occupies the first 106 pages of the book, Ashmole thus giving it pride of place. It is anonymous, but its authorship is revealed by a cipher so easy to decode that no one has been tempted to elevate it into another Bacon–Shakespeare controversy.

As is pointed out by Ashmole, 'From the first word of this Proeme [to the poem], and the Initiall letters of the six following Chapters ... we may collect the Authors Name and place of Residence: For those letters (together with the first line of the seventh Chapter) speak thus,

> Tomas Norton of Briseto,
> A parfet Master ye maie him trowe.'

Ashmole was not quite accurate in his transcription, for, from the text, Tomas should read Tomais and 'call' should have been inserted between 'him' and 'trowe'; but that the author intended the perspicacious reader to identify him as Thomas Norton, a perfect master of the Art and a Bristolian, is indubitable. 'Our author', says Ashmole, 'thus modestly and ingenuously unvailes himselfe; although to the generality of the world he meant to passe unknowne, as appeares by his owne words:

For that I desire not worldly fame,
But your good prayers unknowne shall be my name.'

The cipher had been observed earlier by Maier, who 'came out of Germanie, to live in England; purposely that he might understand our English Tongue, as to Translate Norton's Ordinall into Latin verse, which most judiciously and learnedly he did'. 'Yet (to our shame be it spoken)', continues Ashmole, 'his Entertainment was too course for so deserving a Scholler', and he returned to Germany in the autumn of 1616.

The Thomas Norton of the 'Ordinall' is usually – though, as Nierenstein and Chapman observed in 1932, not reliably – identified with Edward IV's privy councillor of that name, who was a native of Bristol and was probably born in the family mansion that stood on the site later occupied by the sixteenth-century St Peter's Hospital, itself destroyed by enemy action in the 1939–45 war. The Nortons were people of importance in Bristol, Thomas's father being sheriff in 1401 and mayor in 1413; he was also a member of Parliament for Bristol for many years. Thomas himself may have been returned for the borough in 1436; he was well off, owing to commercial activities in which he engaged, and is said to have accompanied the king when he fled to the Continent in 1470. Eight years later a Thomas Norton accused the mayor of Bristol of high treason, and challenged him to a duel in the council-chamber; whether this was the alchemist or not is uncertain, but, if it was, the challenge must have been one of the last acts of his life, for we hear no more of him.

Norton began to study alchemy at an early age, and was soon seeking information about it from adepts. One of his correspondents was George Ripley (p. 187), who, in response to Norton's entreaties, at length wrote a letter inviting him to a personal meeting. Ripley explained that it was necessary that 'Wee speake together, and see face to face', because if he wrote the secrets of alchemy he would be breaking faith with the fraternity of the Art. He ends on a friendly note:

Noe more to you at this present tyde,
But hastily to see me, dispose you to ride.

Norton continues:

> This Letter receiving, I hasted full sore,
> To ride to my Master an hundred miles and more;
> And there Forty dayes continually,
> I learned all the seacrets of Alkimy.

This happened when Norton was twenty-eight years old. Having returned to Bristol he put his newly acquired knowledge to the test of experiment and succeeded in preparing the Great Red Elixir, only to have it stolen from him by a dishonest servant. This so discouraged him that:

> ... remembring the cost, the tyme, and the paine,
> Which I shulde have to begin againe,
> With heavie hearte farewell adieu said I,
> I will noe more of Alkimy.

The fit of depression wore off, however, and he set to work again, this time to prepare the Elixir of Life; but here also he was dogged by misfortune, for though his efforts were crowned with success he was once more robbed of the elixir, this time by a woman who is said on wholly inadequate evidence to have been the wife of William Canynges, a master mason – in those days equivalent to an architect – who rebuilt the beautiful church of St Mary Redcliffe with the profits accruing from the elixir.

Norton took comfort in the thought that he was not the only one to suffer for his devotion to alchemy, and relates the story of an adept named Thomas Daulton, who was a good man, serving God both day and night, and who had such a store of the Red Medicine or elixir that 'never English man had more'. Daulton was leading a peaceful life at an abbey in Gloucestershire when one of Edward IV's courtiers, Thomas Herbert, took him away against his will and brought him before the king. Another of Edward's squires, Sir John Delves, whose chaplain or secretary Daulton had been, knew of these happenings, and although he had been sworn to secrecy revealed to the king that Daulton had once made a thousand pounds' worth of gold in less than half a day – 'as good Goulde as the Royall was'. Daulton reproached Delves for his breach of oath, but Delves excused himself by saying it was for the good of the king and the realm. Daulton then confessed to the king that the possession of the elixir had caused him a great deal of anxiety, and that in the end he had thrown it into a muddy lake communicating with the Severn, so that it was irrevocably lost. His modesty could not restrain him from adding that the gold procurable by means of the elixir would have been sufficient to equip and maintain twenty thousand men on a crusade to the Holy Land.

More in sorrow than in anger, the king said 'Alas Daulton, it was fowly don to spill such a thinge', and ordered the alchemist to make a further supply. Daulton replied that there was no certainty of success, and that he had in fact not made the elixir himself but had obtained it from a canon of Lichfield now dead. The king perceived that nothing was to be gained by taking the matter any further, and dismissed Daulton with a money present of four marks or about £2 13s 4d, which in the circumstances of his disappointment must be considered a generous action. Thomas Herbert was not so easily satisfied. He lay in wait for Daulton, who was lured to Stepney by a false message, and carried him off to Gloucester Castle, thinking by threats to force the alchemist to prepare more of the elixir. Failing to do so, he cast Daulton into prison for nearly four years and at length gave him the choice of revealing the secret or suffering execution. The alchemist was steadfast in his resolve to maintain silence, so the scaffold was prepared and he was led to the block; but when Herbert saw that he was cheerfully ready to die rather than disclose the hermetic mystery he wept in admiration of such courage and resolution and set his prisoner free. Daulton did not long survive his release, and Norton tells us that Herbert died shortly afterwards; Sir John Delves was beheaded after the battle of Tewkesbury in 1471, although he had taken sanctuary in a church and had been given safe conduct.

It was six years after this event, namely in 1477, that Thomas Norton sat down to write 'Of Alkimy the Ordinal, the *Crede Mihi*, the Standard Perpetuall'. Ashmole tells us that, to prepare the text of it that he published in the *Theatrum Chemicum Britannicum*, he compared fourteen manuscript copies, of which one was written on vellum in a very precise and neat hand and was embellished with figures most exquisitely drawn – better than those in Henry VII's copy of the work. This fine manuscript bore the arms of George Nevill, Archbishop of York, and Ashmole believed it might be the original presented by the author, or intended for presentation, to the Archbishop, who died in the same year. This was the same Archbishop to whom Ripley dedicated his 'Marrow of Alchemy'; he had much correspondence with contemporary alchemists.

The 'Ordinall' consists of a 'proheme' and seven chapters, and is written in a lively verse comparable, according to Ascham, to that of the somewhat later poets Wyatt and Surrey. It contains much about alchemists and a good deal about furnaces, but is typically vague about the procedure to be followed in effecting transmutations. Norton indeed professes to speak more clearly than his predecessors, among whom he names Hermes, Rhazes, Geber, Avicenna, Democritus, Morienus, and Roger Bacon, but at the same time emphasizes the esoteric nature of alchemical learning, which he calls a wonderful science and secret philosophy. It can, and should properly, be communicated only by personal teaching, but he says that in the 'Ordinall' he will venture to express it in such a way that the acute and educated may comprehend while laymen will be led astray. Like other alchemists, Norton maintains that alchemy is holy knowledge, revealed from God, and that properly used

> It voydeth vaine Glory, Hope, and also dreade:
> It voydeth Ambitiousnesse, Extorcion, and Excesse:
> It fenceth Adversity that shee doe not oppresse.
> He that thereof hath his full intent,
> Forsaketh Extremities, with Measure is content.

In other words, alchemy though certainly enabling the adept to make gold at will, and thus ease his earthly life, will also preserve him from a wrongful use of money and will ennoble his moral character. The secrecy in which it is wrapped is to prevent abuses, and to discourage those who undertake its study merely from 'appetite of Lucre and Riches'. There are many such, says Norton:

As Popes with Cardinalls of Dignity,
Archbyshopes with Byshopes of high degree,
With Abbots and Priors of Religion,
With Friars, Heremites, and Preests manie one,
And Kings with Princes and Lords great of blood,
For every estate desireth after good;
And Merchaunts also which dwell in the fiere
Of brenning Covetise, have thereto desire;
And Common workemen will not be out-laste,
For as well as Lords they love this noble Crafte;
As Gouldsmiths whome we shulde repreve
For sights in their Craft meveth them to beleeve:
But wonder it is that Wevers deale with such warks,
Free Masons and Tanners with poore Parish Clerks;
Tailors and Glasiers woll not thereof cease,
And eke sely Tinkers will put them in the prease
With greate presumption....

Norton discountenances books of recipes for making the elixir; he says that such receipts should rather be called 'deceipts', and stresses the point that the alchemist should understand the reasons for the operations he carries out:

Nothing is wrought but by his proper Cause:
Wherefore that Practise falleth farr behinde
Wher Knowledge of the cause is not in minde:
Therefore remember ever more wisely,
That you woorke nothing but you knowe howe and whie.

The alchemical aspirant must also learn to distinguish honest workers from charlatans and swindlers:

The trew men search and seeke all alone
In hope to finde our delectable stone,
And for that thei would that no Man shulde have losse,
They prove and seeke all at their owne Coste;
Soe their owne Purses they will not spare,
They make their Coffers thereby full bare,
With greate Patience thei doe proceede,
Trusting only in God to be their speede.

On the other hand, the fraudulent man

. . . walketh frome Towne to Towne,
For the most parte in a threed bare Gowne;
Eever searching with diligent awaite
To winn his praye with some fals deceit
Of swearing and leasing [lying]; such will not cease,
To say how they can Silver plate increase.
And ever they rayle with perjury;
Saying how they can Multiplie
Gold and Silver, and in such wise
With promise thei please the Covetise,
And causeth his minde to be on him sett,
The Falsehood and Covetise be well mett.

The 'multipliers' were alchemists – true or false – who maintained that it was possible to make metals grow and increase, but Norton will have none of this. He says it is true that, in Nature, the subterranean 'vertue Minerall' is gradually converted into metals under the influence of the Sun's rays; but that this happens only in 'certaine places of eligible ground' and that 'few grownds be apt to such generation'. The process cannot be effected by human agency, and the aim of alchemy is the more modest one of transmuting already existing metals, which is possible owing to their 'propinquity of matter'. In support of his view, Norton describes seven images set up by 'Ramon Lull' (p. 126) in a city of Catalonia, 'the trewth to disclose':

Three were good Silver, in shape like Ladies bright;
Everie each of Foure were Gold and did a Knight:
In borders of their Clothing Letters like appeare,
Signifying in Sentence as it sheweth here.
1. Of old Horshoes (said one) I was yre,
Now I am good Silver as good as ye desire.
2. I was (said another) Iron set from the Mine,
But now I am Gould pure perfect and fine.
3. Whilome was I Copper of an old red pann,
Now I am good Silver, said the third woman.
4. The fourth saide, I was Copper growne in the filthy place,
Now am I perfect Gould made by Gods grace.
5. The fift said, I was Silver perfect, through fine,
Now am I perfect Goulde, excellent, better then the prime.

6. I was a Pipe of Leade well nigh two hundred yeare,
And now to all men good Silver I appeare.
7. The seventh said, I Leade am Gould made for a Maistrie,
But trewlie my fellowes are nerer thereto then I.

When Norton comes to deal with the nature of the 'Stone', he puts it in the form of replies to questions asked by an unsuccessful 'labourer in the fire', who for threescore years had experimented with herbs, gums, roots, grass, antimony, arsenic, honey, wax, wine, quicklime, vitriol, and a wide miscellany of other bodies. Having invariably failed to elicit an elixir from these sources, the disappointed man besought Norton to reveal the secret. Norton required much persuasion, but in the end says that for the White work, that is, for preparing silver by transmutation, two constituents go to form the appropriate Stone. The first is a mineral that cannot be bought but has to be prepared by the alchemist himself; it is a subtle earth, dull and reddish-brown, and is known as litharge. Later, presumably after suitable treatment, it turns white, and is then called marcasite. The other constituent is a stone 'gloriouse faier and bright ... glittering with perspecuitie, being of wonderfull Diaphanitie'; it costs twenty shillings an ounce and known as magnetia [*sic*]. When mixed with sal ammoniac and sulphur and heated, these two minerals will yield the White Elixir.

The actual work of transmutation on a 'quantity reasonable' of matter requires the help of eight servants, though for a small quantity four men will suffice. They must be sober, diligent, attentive, obedient, and cleanly, and since the operations must go on day and night continuously they must work in shifts, half being on duty while the other half sleep or go to church. The necessary apparatus is also described; it includes vessels made of clay, stone, glass, and lead. Most important of all is the furnace. Many different degrees of heat are required for different stages of the operation, and Norton truly says that

Olde Men imagined for this Arte
A speciall Furnace for everie parte.

To obviate this cumbrous equipment, he invented a new all-purposes furnace, in which

Threescore degrees divers ye maie gett,
For threescore warkes, and everie-ech of divers Heate,
Within that Furnace, to serve your desire,
And all thi served with one litle Fier,
Which of a Foote square onlie shalbe.

It seems likely that the principle of this multiple furnace was an arrangement of dampers or 'stopples', devices apparently unknown before Norton's introduction of them. He says:

Consider your Stoples, and lerne well this,
The more is the Stople the lesse is the Heate,
By manifould Stoples Degrees ye maie gett.

For carrying out the work, says Norton, a suitable place must be found,

For manie things woll wonderous doe
In some Places and elsewhere not soe.

For some stages the surroundings should be dry, for others moist and cold; for some, a dim light is needed, for others the light cannot be too bright. In all cases, however, wind is harmful. Similarly, the operations should be carried out with due regard to astrological considerations; figures given in the 'Ordinall' as printed in the *Theatrum Chemicum Britannicum* indicate that in four successive stages the Sun should be in *Sagittarius* and the Moon in *Aries*, the Sun in *Libra* and the Moon in *Virgo*, the Sun in *Virgo* and the Moon in *Libra*, and finally the Sun and Moon both in *Leo* to effect perfect union (plate 9).

Another alchemical poem ascribed to Thomas Norton is as follows:

Take Earth of Earth, Earthes brother
and water of Earth that is an other
and fyer of Earth that beareth the pryce
and of that Earth looke thow be wyse.
This is the true Elixir for to make,
Earth owt of Earth looke thow take;
pure, suttle, fayer and good,
and then take the water of the wood
there as Crystall shyning bright
and doo them together anone right,

three dayes then let them lye
and then departe them privily and sly
then shall yt be bright shyning
and in that water a Soule raigning,
Invisible, hyd, and not sene,
a marveilous matter yt is to meane.
Then depart them by destilling
and thow shallt see an Earth appearing
heavy as Mettall should yt be
in the which is hyd great privitie.
Destill that in a grene hewe
three dayes during well and trewe.
And doo him then in a Bodie of glas
in the which never any woork was.
In a furnace hee must be set
and doo on him a good Lembik
and draw from him a water clere
the which water hath no pere
and then make thy fyer stronger
and thereon contynue thy glas longer
and then truly shall come a fyer
red as Blud and of great yre
and after that an Earth leave there shall
the which is called the Mother of all.
Then into Purgatorie she must be doo
and have the paynes that longes thereto
till she be brighter then the Sonne
for then have ye all the Maistrie wonne.
And that shall be within howers three
and that will shew a great privitie
then doo her in a fayer glas
with some of the water that hers was
and into a furnace doo her againe
till she have drunk her water certaine
and after that water geve her Blud
that was her own pure and good,
And when she hath drunk all the fyer
she will wax strong and of great yre
Then take Meate and Mylk theretoo
and fede that Childe as thow shouldst doo
tyll hee be growen of his full age
then shall hee be strong and of great courage

and turn all bodies that lawfull be
to his power and dignitie
And this is the making of our Stone
the truth I have tolde you everichone.

THOMAS CHARNOCK

Thomas Charnock was born in 1526, or, according to his own statement, in 1524, at Faversham in Kent, that is, about fifty years after Norton had written the 'Ordinall'. A good deal is now known of his life, thanks to the researches of Sherwood Taylor, on which the following account is largely based. Charnock describes himself as an 'unlettered Scholar', but that probably means merely that he was not proficient in Latin and Greek; he could certainly read and write, and was a fluent versifier. In his early twenties he travelled all over England in search of alchemical knowledge, settling for a time in Oxford. Here he became friendly with 'James S., a spiritual man', who lived at Salisbury and was an accomplished alchemist; his name may have been Sauler. 'J. S.' made Charnock his laboratory assistant, and dying in 1554 bequeathed to him the secret of the philosophers' stone. Unluckily his apparatus was destroyed by fire on the following New Year's Day, and he seems to have been careless enough not to commit the secret either to memory or to paper, for he was forced to learn it again from the last Prior of Bath, William Holway (p. 188). When Bath Abbey was surrendered to the Crown, in 1525, Holway received a pension of £80 a year, but that did not compensate him for a grievous mischance; he had, so he told Charnock several years later, possessed the Red Elixir but had hidden it at the time of the dissolution of the Abbey. When, sometime afterwards, he went to look for it in the wall where he had secreted it, he could not find it, and was so overcome by grief that he temporarily lost his reason.

However, according to Ashmole the elixir had not been stolen, for when workmen subsequently pulled down some stone-work of the Abbey,

there was a Glasse found in a Wall full of Red Tincture, which being flung away to a dunghill, forthwith coloured it, exceeding red. This dunghill (or Rubish) was afterwards fetched away by Boate by Bathwicke men, and layd in Bathwicke field, and in the places where it was spread, for a long tyme after, the Corne grew wonderfully ranke, thick, and high: insomuch as it was there look'd upon as a wonder. This Belcher and Foster (2 Shoomakers of Bath, who dyed about 20 yeares since) can very well remember; as also one called Old Anthony, a Butcher who dyed about 12 yeares since.

This Relacon I recd: from Mr Rich: Wakeman Towne Clearke of Bath; (who hath often heerd the said Old Anthony tell this story) in Michaelmas Tearme 1651.

It was when Holway was a blind old man that Charnock met him, probably shortly after the catastrophe that had destroyed his first 'Work', and when Holway had sworn both of them to secrecy he supplied Charnock with the essential knowledge. Charnock thereupon set experiments in train again, first with a servant to help him and then by himself, and after several months he believed that he was near success. But he was doomed to disappointment, for the Duke of Guise was attacking Calais, the last English possession in France, and in 1557 Charnock was called up for the army at the critical moment. This so infuriated him that in a rage he seized a hatchet and smashed his apparatus to bits. He describes the event as follows:

> Then a gentleman that ought me great mallice,
> Caused me to be pressed to goe to serve at Callys:
> When I saw there was none other boote
> But that I must goe spight of my heart roote;
> In my fury I tooke a Hatchet in my hand,
> And brake all my work whereas it did stand.

Ashmole comments that Charnock must have been a poor man, for otherwise he could have bought himself off from military service; but the pursuit of practical alchemy was expensive, and if Charnock had thought himself so far along the road to the Stone he might well have been lavish in expenditure.

Calais fell in 1558, so presumably Charnock was in England very quickly and able to resume his search. In 1562 he married Agnes Norden, of Stockland Bristol, Somerset, and in the following year had a son who died in infancy; a daughter also is mentioned. From Stockland he moved to Combwich, near Bridgwater, where he set up a laboratory and pursued his experiments until his death in April 1581; he was buried in the neighbouring village of Otterhampton.

About a hundred years later a parchment roll about six feet long and nine inches wide was discovered in the wall of a room in Charnock's house at Combwich. It dealt with alchemy, and a clergyman named Andrew Pascal, having heard of the discovery, went there to make inquiries. In a letter to John Aubrey, F.R.S., the antiquary and a friend of Ashmole, Pascal writes:

> I was also since my last at Mr Charnocks house in Comage [Combwich], where the Roll was found, and saw the place where it was hid. I saw a little roome, and the contrivance he had for keeping his worke, and found it ingeniously ordered, so as to prevent a like accident to that which befel him New Years day 1555, and this pretty place joyning as a closet to his Chamber was to make a Servant needles[s], and the work of giving attendance more easy to himself. I have also a little Iron Instrument found there which he made use of about his Fire. I saw on the dore of his little Athanor-room (if I may so call it) drawne by his owne hand, with course colours and worke, but ingeniously, an Embleme of the Worke, at which I gave some guesses, and soe about the walls in his Chamber, I think there was in all 5 panes of his worke, all somewhat differing from each other, some very obscure and almost worne out. They told me that people had been unwilling to dwell in that house, because reputed troublesome, I presume from some traditionall stories of this person, who was looked on by his Neighbours as no better than a Conjuror. As I was taking Horse to come home from this pleasant entertainment, I see a pretty ancient man come forth of the next door. I asked him how long he had lived there, finding that it was the place of his birth, I enquired of him, if he had ever heard anything of that Mr Charnock. He told me that he had heard his Mother (who died about 12 or 14 years since and was 80 years of age at her decease) often speake of him. That he kept a fire in, divers years; that his daughter lived with him, that once he was gon forth, and by her neglect (whome he trusted it with in his absence) the fire went out, and so all his work was lost.

This undependable daughter was named Bridget; she married one Thatcher of Stockland in 1587.

Charnock says that not until 1579 did he succeed in preparing the Stone, after twenty-four years of labour, but it is difficult to tell whether he honestly believed that he had achieved what so many had failed to accomplish. At least in his younger days he was more than a little sceptical about the pretensions of the adepts, as may be gathered from his 'Breviary of Naturall Philosophy' written while he was at Calais in 1557. Addressing Holway, he says:

> ... I met with two Philosophers in Calais towne.
> The cause you know why that thither I far'd,
> Now there I found a kinsman of Queene Maries guard,
> Whom I was glad to see, and hee me againe,
> But what Joy was made betwixt us twaine
> At our meeting, I leave out for this time;
> And to the best Inne wee went to drink wine,
> And there for to lodge, and to make good cheare,
> For wee had not seene each other in many a yeare.
> Wee had our chamber appointed anone,
> There for to supp most quietly alone:
> But so it chanced ere the cloth were laid
> In came our host, and to my kinsman said,
> Mr Charnock, pleaseth you to understand,
> That here be two yeomen come from Ingland
> Who would be glad to have your company:
> My kinsman said with countenance merrily,
> I pray you mine host let them come nier,
> Good fellows fellowshipp is greatly my desire.

It turned out that the two 'yeomen from England' were students of alchemy, and had with them a satchel containing

> ... Sulphur and Mercury the tincture to all metall,
> With all the 7 salts which to our Science must goe,
> With oiles and corrosives, which the ignorant doe not know.

Charnock did not reveal that he himself was an adept, but with his kinsman listened to the yeomen's talk after they had gone to bed. They were full of hope, and discussed what they would do with the alchemical gold when they had made it:

They would live as Princes all their lives after:
And home they would to Winchester Citie
Where they were borne, and do workes of pitty,
And also re-edify the Walls and Castle, they saies,
And make it beautiful as in King Arthurs daies:
And bring again the old River from Southampton,
That lighters and boates may goe and come
With any kinde of merchandise, fewell, or ware,
And of all these charges for gold they had no care
And then a Market crosse in that Citie they would up set,
The patterne whereof in Cheapside should be fett [fetched],
Wherein their Images should stand both together,
Well guilded with gold, a memoriall for ever.

The next day Charnock warned them that alchemy would soon empty their coffer, and that they would not get the elixir from their sulphur and mercury – unless they had also brought the Man in the Moon in their satchel! The yeomen realized that Charnock knew much more about the Art than they did themselves, and offered to entertain him and give him a gold Portegue (=£4) if he would teach them the substance of the matter. As he declined, they continued along their own lines, spending much money but at length summoning Charnock to see a fair crystal stone which they said they had made and believed to be the White Elixir. But, says Charnock,

I promise you truely it was worth right nought,
As I proved to their faces ere that I went;
And then they began themselves to repent,
And said they had spent a hundred mark and more,
Yea I said then, I told you this before,
So cursed they the Science and said it was not true,
But what became of them after, I knew not, for I bid them adieu.
But of anything that I could heare, by word or by letter,
Winchester Citie for their Philosophy was ne're the better.

In 1566 he wrote a book entitled 'A Booke of Philosophie', which he dedicated to Queen Elizabeth and of which he delivered a copy to her chief secretary, William Cecil, first Baron Burghley. Charnock says that the book was placed in the Queen's library; he had rather rashly written in it that on pain of losing his head he would do the thing 'that all this realm should not do agayne', that is, make gold by alchemical means. It was perhaps fortunate for him that the book was laid aside for a time; Elizabeth's cupidity might otherwise have commanded him to justify his boast. The book vanished later, and one wonders whether it came into the hands of the 'Wizard Earl' (p. 58), who, early in the succeeding reign, practised alchemy in the Tower of London.

EDWARD KELLY AND JOHN DEE

In *The Alchemist*, Ben Jonson refers to an alchemist named Kelly, an egregious scoundrel whose association with the learned astrologer and mathematician John Dee fills one of the less reputable pages of alchemical history. Dee, who was of Welsh descent, was born in London on 13 July 1527 and entered St John's College, Cambridge, in the Michaelmas Term of 1542. He took his B.A. in 1545, and in the following year was elected to a fellowship at his college. However, 1546 saw the foundation of Trinity College by Henry VIII, and Dee transferred thither as one of the original fellows. Here he taught Greek, and enlivened his teaching by producing the comedy of 'Peace' by Aristophanes. In one of the scenes of this play, a man carrying a basket of provisions is transported to heaven on the back of a huge beetle, and Dee managed the stage-effects so cunningly that 'many vain reports spread abroad of the means how that was effected'; the consensus of opinion was that Dee had invoked the aid of magic, and he thus earned a reputation that many of his later doings served to confirm. The years 1547 to 1551 he mostly spent abroad in the Low Countries and Paris, though he returned for a time to England in 1548 to take his M.A. – the highest degree ever conferred upon him, in spite of his usual description as 'Doctor'.

In 1551 Edward VI granted him an annual pension of a hundred crowns, and two years later he became rector of Upton-on-Severn in Worcestershire. Here he continued his mathematical studies, but he declined an offer of a mathematical lectureship at Oxford. His troubles began after the accession of Queen Mary I, for some correspondence he had with Princess Elizabeth's *entourage* brought him under suspicion of having attempted to take the Queen's life by magic. He was arrested, his London apartments were searched and sealed, and he was brought before the Star Chamber. After trial, he was found not guilty of treason, but his religious views were suspected of being heretical and until 1555 he was confined in the Bishop of London's prison. He was then released on promise of good conduct.

On the accession of Elizabeth I in 1558 he was taken into the service of the court, the Queen saying, 'Where my brother hath given him a crown, I will give him a noble.' He was called upon to calculate astrologically a suitable day for the coronation, and Elizabeth was so well satisfied with the result that she promised to make him master of the hospital of St Katharine-by-the-Tower. The promise was not fulfilled, however, and after waiting some time in vain, Dee set off on travels through Germany and Italy, and also voyaged to St Helena. Secretary Cecil expressed the view that Dee's time abroad had been well spent in the pursuit of knowledge, and when he returned to England in 1564 he found himself still in Elizabeth's favour – though, like many others, he soon found that royal promises of reward were not always to be relied upon. His benefices brought enough income for him to buy a riverside house at Mortlake, and there he lived for several years in study and in amassing an extensive library of curious books and manuscripts. His interests were tending more and more to the occult, and he acquired a fine scrying-glass for crystal-gazing; this glass, a globe of polished smoky quartz, is still preserved in the British Museum, as are various waxen tablets, covered with arcane figures, used by Dee in his magical rites.

News of this crystal reached the Queen, who, accompanied by a number of courtiers, made a special journey to Mortlake to see it; but hearing that Dee's wife had just been buried she would not go into the house and the glass was brought out for her inspection. Shortly afterwards a comet appeared in the heavens and Dee was summoned to Windsor to explain its significance, which took him three days; and on another occasion his presence was urgently commanded in order to prevent any ill befalling the Queen from a waxen image of her Majesty found, in Lincoln's

Inn Fields, with a pin thrust through its breast. His efforts must have been successful, for the Queen survived the event by a quarter of a century.

By 1580 Dee had become fully engrossed in magic and alchemy, and believed himself to be in communication with angels and spirits. He dressed the part, and must have made a striking figure. John Aubrey says that he was tall and slender, with a very fair, clear, and sanguine complexion and a long beard as white as milk. He wore a gown like an artist's, with hanging sleeves and a slit. He was a great peacemaker; if any of his friends quarrelled he would never be content until he had reconciled them. He kept a great many stills going, and children dreaded him because they thought he was a wizard.

Besides the globe, Dee had a disk for scrying; this was made of polished cannel coal. Both had to be suitably manipulated, and spirits would then appear either in the globe or disk or disembodied in the room. Only one person at a time could see the spirits, so Dee engaged a scryer and himself took pen and paper to write down the spirit messages. The first scryer seems to have been honest enough, but unfortunately Dee now became acquainted with Kelly. Probably Kelly had heard of Dee and thought him a likely dupe; another opinion is that Kelly was sent as a spy to discover the nature of Dee's magical practices.

Edward Kelly or Kelley, *alias* Talbot, was born at Worcester in 1555 and was apprenticed to an apothecary, but he went to Oxford for a time, leaving the university abruptly without taking a degree. He next appears in London, where he had gained a reputation as a fraudulent lawyer, and then at Lancaster, where about 1580 he had his ears cropped in the pillory for forging or coining; he had, moreover, got into trouble by digging up a corpse at Walton-le-Dale for purposes of necromancy. On 10 March 1582 this thorough-paced rogue called upon Dee at Mortlake, wearing a black skull cap to conceal his lack of ears. Assuming a solemn manner, he announced that he was a serious student of the occult and had come to Mortlake in the hope that so eminent a practitioner as Dr John Dee would favour him with some instruction. Dee at first would not admit to the possession of any knowledge of magic, but Kelly's suave tongue had its effect and Dee at length produced his crystal. Kelly must clearly have thought out his plan of action well in advance, and had learned enough of the jargon to impress the gentle and unsuspicious doctor. The crystal was set in place, prayers were offered, and Kelly assumed the office of scryer.

It is perhaps not surprising that great success was achieved. A spirit appeared, announced its name as Uriel, gave directions for the invocation of other spirits, and ordered that an evil spirit named Lundrumguffa, who was inimical to Dee, should be dismissed. Uriel was particularly emphatic that Dee should engage Kelly as regular scryer and that the pair should always work together. Dee had by this time married again, but his wife apparently made no objection to Kelly as a member of the household, and he was engaged at an annual salary of £50. When he wanted a rise, he had only to threaten to leave, for Dee placed perfect confidence in him and believed all the revelations he received from the spirits. Alchemical research went on at the same time as the crystal-gazing, and no doubt much of Dee's gold was transmuted into a lining for Kelly's pockets.

In July 1583 Dee and Kelly were visited by the Earl of Leicester and a Bohemian friend of his, a nobleman named Albert Laski. Laski had fallen on evil times and hoped to restore his fortunes by discovering the philosophers' stone. After some discussion at dinner – for which Elizabeth had provided the cash – it was agreed that Laski should stay on at Mortlake and collaborate with Dee and Kelly; but the experiments proved costly, money was running short, and Laski therefore suggested that the three of them should move to his castle near Cracow, where they arrived in February 1584. Some months passed in fruitless research, and the disillusioned Laski then decanted his alchemical friends on to the Emperor Rudolf II at Prague. One interview was enough for Rudolf, so Dee and Kelly obtained an introduction to King Stephen of Poland, with whom, however, they fared no better. Stephen attended one of the séances, detected imposture, and returned the pair to Prague. Here the papal nuncio obtained a decree ordering them to leave the city within six days, and they set off hurriedly for Erfurt, only to be turned back by the municipal authorities when they arrived. They were able to stay for a time at Hesse-Cassel, and then to their relief were invited by Count Rosenberg to his castle at Tribau in Bohemia, where they resumed their study of the occult.

Some time before they left England, Dee and Kelly had journeyed to Somerset, where, in the ruins of Glastonbury Abbey, Kelly had the exceeding good fortune to unearth a supply of the philosophers' stone prepared by no less a person than St Dunstan himself, and therefore of unquestionable authenticity. By even greater good fortune, Kelly still had some of this powder when he and Dee arrived at Tribau, and was thus able to demonstrate a transmutation before the Count, one small grain converting an ounce and a quarter of mercury into nearly an ounce of gold. Ashmole relates that Kelly also transmuted into gold a piece of metal cut out of a warming-pan, and sent both the gold and the warming-pan to Queen Elizabeth as ocular proof.

But that was not the limit of Kelly's effrontery. In April 1587, while still at Tribau, he saw in the crystal a naked woman who directed that in future the scryer and his master should have their wives in common. Even Dee demurred at this, but Kelly emphasized the impiety of refusing to obey the command of the spirit, and in the end Dee agreed. He himself has left a record of the transaction: 'On Sunday, the third of May, Anno 1587, I, John Dee, Edward Kelley, and our two wives [Jane Dee and Joan Kelley], covenanted with God, and subscribed the same, for indissoluble and inviolable unities, charity, and friendship keeping between us four; and all things between us to be common, as God by sundry means willed us to do.' The outcome was as might be expected; bitter quarrels broke out and early in 1588 Kelly left Tribau for Prague, where the Emperor Rudolf soon cast him into prison. He was released after four years, but Rudolf imprisoned him again about a year later, and he was killed while trying to escape.

Dee returned to England in 1589 and was received by the Queen at Richmond; she made him a present of 100 marks and awarded him a pension of £200 a year. Later he was given the wardenship of Manchester College, but he was not happy in this post and his health was beginning to fail. He returned to Mortlake in 1604, and died there in 1608. He was buried in Mortlake church. His son Arthur (p. 214), also an alchemist, was physician to the Tsar of Russia and later to Charles I.

SIR KENELM DIGBY

A part-time alchemist who cut a dashing figure in the seventeenth century was Sir Kenelm Digby, of whom John Aubrey says:

> [He] was held to be the most accomplished cavalier of his time. He was such a goodly handsome person, gigantique and great voice, and had so graceful elocution and noble addresse, etc., that had he been drop't out of the clowdes in any part of the world, he would have made himselfe respected. He was a person of very extraordinary strength. I remember one at Sherborne protested to us, that as he, being a midling man, being sett in a chaire, Sir Kenelme tooke up him, chaire and all, with one arme. He was of undaunted courage, yet not apt in the least to give offence. He was a great traveller, and understood 10 or 12 languages. He was not only master of a good and graceful judicious stile, but he also wrote a delicate hand, both fast-hand and Roman. He was well versed in all kinds of learning. Since the restauration of Charles II, he lived in the last faire house westward in the north portice of Convent garden, where my lord Denzill Hollis lived since. He had a laboratory there. I think he dyed in this house.

The *preux chevalier* was the elder son of Sir Everard Digby, and born in 1603, three years before his father was executed for complicity in the Gunpowder Plot. An astrological nativity scheme, drawn up by Digby himself, says that his birth took place, 'according to the English account, the 11 of July betweene five and six of the clocke in the morning'. Though the Crown had confiscated most of his father's estate, Digby was left fairly well off, and in 1618 entered Worcester College (then Gloucester Hall), Oxford, where his tutor was Thomas Allen, a mathematician and occultist. He came down in 1620, however, without taking a degree; he had fallen in love with a very beautiful and accomplished girl, Venetia Stanley, who had been his playmate in infancy, but his mother did not approve of the match so recalled him from Oxford and sent him abroad. He went first to Paris, but an outbreak of plague occurred not long after his arrival and he moved to Angers. Here, we are told, he met Marie de Médicis, the queen mother, at a masked ball. Alarmed at the immodest suggestions she made to him, he spread a report of his death and fled to Italy, where he spent two years at Florence. He was then invited to Madrid by the English ambassador there, who was a kinsman, and while in that city he met Prince Charles (afterwards Charles I) and the Duke of Buckingham. Charles took a fancy to the young man and admitted him to his household, and Digby voyaged back to England with the prince in 1623. He was knighted a few days after arrival, and relates that James I was so nervous at the sight of the naked sword that, but for Buckingham's quick intervention, the point would have been thrust into his eye.

Digby then sought Venetia, but unluckily, though the story of his death had reached her, the letters he wrote explaining what had really happened had gone astray; in the meantime she had become the mistress of Sir Edward Sackville. However, when Digby arrived they found that their mutual affection was as strong as ever, and they were married early in 1625, the marriage proving a very happy one.

Two years later Digby resolved upon a privateering expedition in the Mediterranean, to improve the family fortunes. Though the king's favourite, Buckingham, was friendly enough to Digby himself, he was at bitter enmity with Digby's kinsman the Earl of Bristol, and Digby saw that there was little chance of preferment at home. The king shilly-shallied when asked for a royal commission for the expedition, so Digby took out letters of marque and set off from Deal just before Christmas, 1627. His ships were the 'Eagle', of 400 tons, and the 'George and Elizabeth', of 250 tons. Arriving in the neighbourhood of Gibraltar in the middle of January, he took several Flemish and Spanish prizes and then sailed on into the Mediterranean. Here he had to anchor off Algiers for five or six weeks, owing to illness among the crews, but on 30 March he captured a rich Dutch ship near Majorca. Still sailing east, he arrived at Scanderoon (Alexandretta), on the coast of Syria, on 10 June, and the next day gave battle to the French and Venetian ships lying in the harbour. Here is his own account of what happened:

In the forenoone my boate came back to me, who brought me certaine news that in the roade were 4 French vessels, whereof one was come in but a day before, and had still one hundred thousand reals of eight [£20,000] aboard her; that withall there were 2 English shippes, 2 Venice galliegrosses, and 2 of their galliones. I stood in with the roade as fast as I could, but before, I sent my sattia boat [pinnace] with letters to the Venetian generall and the English captaines, to acquaint them who I was, contriving it so that my letters should be delivered even as I came within shott. The Venetian generall treated my men ill, and sent me word he would sinke my shippes if I went not out of the roade. He did his best, and shott att my flagge, but after I had endured 8 shott from him patiently, and saluted him with gonnes, I then fell upon his vessels with all my might. It continued a cruell fight for about 3 houres. It was most part calme, else I had offended him much more. Towards night the wind freshed; then I prepared to bord the galliones, and so meant to stemme the galeazzes, for I could easily gett the wind of them, having much maimed their oares, and they being so frighted (as it appeared by their working and the issue) that they lost all their advantages. Then the generall sent to me beseeching peace, and acknowledging his error in a very abject manner, having hoisted his yards to be gone out of the roade in case I refused it. I had taken then all but one (French), who was runne aground.

News of Digby's capture of the 'huge galeazzoes of St Mark' was received with enthusiasm in England, whither he returned in February 1628 and met with commendation from the king – though the government prudently disavowed his piracy. The expedition must have put him in funds, for he does not appear to have had any financial difficulties afterwards; but he suffered a heavy blow in 1633, when to his profound grief his wife died. Aubrey tells us that he then retired to Gresham College, London,

where he diverted himselfe with his chymistry, and the professors' good conversation. He wore there a long mourning cloake, a high crowned hatt, his beard unshorne, look't like a hermite, as signes of sorrow for his beloved wife, to whose memory he erected a sumptuouse monument [in Christ Church, Newgate] now quite destroyed by the great conflagration [1666]. He stayed at the colledge two or 3 years.

Digby's royalist leanings and Roman Catholic sympathies brought him under the suspicion of the Long Parliament, and he was summoned to the bar in 1640, the Commons afterwards petitioning the king to remove him from the privy council as a Popish recusant. He went to France, and while he was there a Frenchman, Mount le Ros, was rash enough to offer an insult to Charles I in his presence. Digby at once challenged le Ros to a duel, and 'in foure bouts hee runne his rapier into the French Lords brest till it came out of his throate again, which so soone as he had done, away hee fled to the court of France and made all knowne to the king thereof, who said the proudest Lord in France should not dare to revile his brother king'. Nevertheless Digby thought it wise to return to England, but on his arrival he was arrested by order of the House of Commons and confined in an inn called 'The Three Tobacco Pipes', near Charing Cross; here, according to a fellow-prisoner, Sir Roger Twysden, his charming conversation made the prison 'a place of delight'. He was later transferred to Winchester House, where he whiled away the time in writing and in practising chemistry, making artificial gems. It was also while in prison that he received from the Earl of Dorset an advance copy of Sir Thomas Browne's 'Religio Medici', which interested him so much that he stayed up all night to read it, then writing a review of it from the Roman Catholic point of view. Sir Thomas was not pleased about it and protested, but the review was already in print.

Digby was released in 1643, through the intervention of Anne of Austria and on condition that he left immediately for France and gave an undertaking not to return without the leave of Parliament. This he obtained in 1653 and took advantage of it to spend a year or so in England from 1654 to 1655, when he went back to the Continent and remained there till the Restoration in 1660. He died on 11 June 1665; his library was still in Paris, and the Earl of Bristol repurchased it for 10,000 crowns.

While at Gresham College, Digby had as assistant one Georg Hartmann, anglicized to George Hartman, who after his master's death published 'A Choice Collection of rare chymical secrets and experiments in philosophy. As also rare and unheard-of medicines, menstruums, and alkahests: with the true secret of volatilizing the fixt salt of tarter. Collected and experimented by the honorable and truly learned Sir Kenelm Digby, Kt, Chancellour to Her Majesty the Queen-Mother. Hitherto kept secret since his decease, but now published for the good and benefit of the publick.'

The 'secrets' are partly medical and partly alchemical, and are not of any particular interest. Digby believed that by heating silver amalgam with red precipitate (mercuric oxide) for three weeks a yellow powder could be obtained which, on fusion with borax, would yield the original silver together with a mass of pure gold equal in weight to the mercury taken; and the other recipes are on the same level of veracity. Digby was in fact incapable of telling the truth about his experiments – though, as Lady Fanshawe remarked, he was otherwise 'a person of excellent parts and a very fine-bred gentleman'. In his diary, John Evelyn says that he visited Digby in Paris in 1651 and had much discourse with him on chemical matters.

I showed him a particular way of extracting oil of sulphur, and he gave me a certain powder with which he affirmed that he had fixed mercury before the late king. He advised me to try and digest a little better, and he gave me water which he said was only rain-water of the autumnal equinox, exceedingly rectified, very volatile; it had a taste of strong vitriolic and smelt like aqua-fortis. He intended it for a dissolvement of calx of gold; but the truth is, Sir Kenelm was an arrant mountebank.

He had a cure for toothache, which consisted in scratching the gums near the aching tooth with an iron nail and then hammering the blood-covered nail into a wooden beam; but his most sublime remedy was a 'Powder of Sympathy' or weapon-salve, of which he said that he had learnt the secret from a Carmelite who had travelled in the East and whom he met at Florence in 1622.

This remarkable powder acted at a distance; it was not to be applied to a wound but to a bandage stained with blood from the wound. Digby says that he first employed it to cure James

Howell (1594?–1666), a Welshman and a prolific author, of a cut in the hand while they were both in Madrid; and adds that James I and Dr Mayerne were greatly impressed by its efficacy. The powder, which had a great vogue for a time, was nothing but green vitriol (ferrous sulphate) but Digby claimed that a current of particles of the powder and blood somehow found its way to the wound and performed its cure. A recent commentator aptly says that 'any success the sympathy cure had was due to the innate ability of wounds to heal themselves when kept clean and uninfected; and wounds left alone had a better chance of healing in many cases than if they had been treated in the insanitary manner of those times'.

Yet in spite of his credulity and untruthfulness Digby was esteemed highly by such famous scientists of the day as Gilbert, Harvey (who discovered the circulation of the blood), and Descartes, and was one of the original Fellows of the Royal Society.

Besides Georg Hartmann, Digby had another operator to help him in his experiments at Gresham; this was a Hungarian whom he calls Hans Hunneades and who is to be identified with Johannes Banfi Hunyades, an alchemist born at Baia Mare about 1576. He came to London in 1632 or 1633 and was professor at Gresham College, where his studies were mathematics and alchemy; but he was also interested in technical chemistry and seems to have worked on a fairly large scale, probably preparing substances for use in medicine. When he began to grow old, he felt a desire to return to his native country, and told a friend, regretfully, that only his years would prevent him from acquiring a fortune there by working up the refuse of the old Roman gold-mines. But he died before he could get there, as we learn from a letter written by Sir Thomas Browne to Ashmole concerning Dr Arthur Dee (p. 209):

> Dr Arthur Dee was a young man when he saw this Projection made in Bohemia but he was soe inflamed therewith, that he fell really upon that Study, & read not much in all his lyfe but Bookes of that subject, and two yeares before his Death, contracted with one Huniades or Hans Hungar in London, to be his operator. This Hans Hungar having lived long in London, & growing in yeares, resolved to retourne into Hungary, he went first to Amsterdam, where he was

to remaine ten weekes, till Dr Arthur Dee came unto him. The Dr to my knowledge, was serious in his buisnies, & had provided all in readines to goe, but suddainely he heard that Hans Hungar was dead.

As Dee died in 1651, it may be presumed that Hunyades died in 1650. A portrait shows him holding alchemical apparatus.

10

SCOTTISH ALCHEMISTS

SCOTLAND had a very early connexion with alchemy in the person of Michael Scot, who lived from about 1175 to about 1235 and was one of the court astrologers to the Emperor Frederick II of Sicily (1194–1250). Frederick was a great patron of learning and himself a scholar; his book on falconry contains numerous facts about the anatomy of birds that had not previously been recorded. His interest in animals led him to collect a menagerie, which he took with him when he made his long journey over the Alps to Aachen for his coronation in 1215; it included not only falcons but owls, monkeys, lions, panthers, leopards, camels, dromedaries, elephants, and the first giraffe to be seen in Europe. He founded the university of Naples, and at his court surrounded himself with Christian, Muslim, and Jewish scholars.

Into this favourable atmosphere for learning, Michael Scot came about 1227. Of his earlier life very little is definitely known. It is possible that he belonged to the Scot family of Balwearie, near Kirkcaldy in Fife, though more probably he came from the border country, where his name is still familiar as Auld Michael the magician. He seems to have studied at Oxford and in Paris, afterwards going on to Bologna and, some say, to Palermo, where for a short time he was in the service of Don Philip, one of Frederick's officials. We next hear of him at Toledo, where, on the testimony of Roger Bacon, he gained a knowledge of Arabic. He took holy orders, and Pope Honorius III (1216–27) in 1224 nominated him archbishop of Cashel in Ireland – to the annoyance of the canons, who wanted the office to be given to the bishop of Cork. However, Michael solved the difficulty by declining the archbishopric on the grounds that he did not understand the Irish language, and was given a benefice in Italy. In 1227, as we have seen, he returned to the court of Frederick II, this time as official astrologer to the Emperor himself.

While in Sicily he made many translations and also wrote his principal work – a general introduction to astrology in four parts – together with at least two works on alchemy, the 'Magistery of the Art of Alchemy' and the 'Lesser Magistery'. In these two books there is evidence that he carried out alchemical experiments, sometimes with Jewish and Muslim collaborators, evidence supported by his own statement in another place, where he says, 'I, Michael Scot, have experienced this many times.' His alchemy consists of descriptions of alums, salts, vitriols, spirits, and many recipes for transmutation. He used a wide variety of substances, including minerals and herbs from Calabria, India, and Alexandria, and such odd ingredients as the dust of moles, the blood of an owl, and opium. According to one recipe quoted by Lynn Thorndike, 'five toads are shut up in a vessel and made to drink the juices of various herbs with vinegar as the first step in the preparation of a marvellous powder for the purposes of transmutation'.

It is little wonder that Michael Scot soon acquired great fame as a wizard or magician, so much so as to be placed by Dante in the *Inferno*:

Quell' altro, che ne' fianchi è così poco
Michele Scotto fu, che veramente
delle magiche frode seppe il gioco
That other there, whose ribs fill scanty space,
Was Michael Scot, who truly full well knew
Of magical deceits the illusive grace.

Scot also figures in Boccaccio's *Decameron* (eighth day, ninth novel). I quote from Richard Aldington's translation:

You must know then, sweet Master, that not long ago there dwelt in this city [Florence] a great master of black magic named Michael Scott (because he came from Scotland), who was greatly honoured by many gentlemen, of whom few are now alive. When he was about to depart he yielded to their earnest entreaties and left behind two competent disciples, ordering them always to be ready to carry out the desires of the gentlemen who had entertained him.

These two served the gentlemen aforesaid in certain love affairs and other slight matters. Finding that they liked the city and the manner of life in it, they determined to remain here permanently and formed close friendships with certain men, not considering whether they were gentle or common, poor or rich, but only whether they were men of their own sort. To please their friends they instituted a club of about twenty-five men who were to meet twice a month in some place determined by them. When they met, each one declared what he needed, and the two magicians quickly provided it that night –

even were it a thousand or two thousand gold florins.

Scot was said to have foretold that the death of the Emperor would occur at Florence, and Frederick therefore refused for the future to enter that town, or even Faenza; but he met his fate at Firenzuola or 'Little Florence'. The magician also flew through the air on a demon horse and sailed the sea in a demon ship. When he was sent as an envoy from Scotland to the king of France, the first stamp of his jet-black steed in Paris rang the bells of Notre-Dame, the second threw down the palace towers, and, to avert the third, the king granted all he asked. Finally Scot predicted the manner of his own death, which he foresaw would happen by the fall of a small stone on to his head. He took the precaution to wear a cerebrerium or steel helmet, but at church one day had to remove it at the elevation of the Host. At that very moment a stone about two ounces in weight fell from the roof of the church and killed him. Scottish tradition has it that his body was brought back to Scotland and interred with his books of magic in Melrose Abbey, a tradition adopted by Sir Walter Scott in his 'Lay of the Last Minstrel'.

Alchemy in Scotland some three hundred years later has been described by John Read. James IV (1473–1513) whose marriage to Margaret, daughter of Henry VII, ultimately led to the succession of the Stuarts to the English throne, was an intelligent prince interested in medicine, surgery, physiology, alchemy, and many other subjects. He made the historic experiment of sending two babies to live with a dumb woman on Inchkeith island in the Firth of Forth in order to discover what language they would speak when they grew up. The chronicler says that he had no first-hand-information on the result, but that it was reported that the children spoke good Hebrew. James practised medicine and surgery, generously paying his patients for allowing him the opportunity to do so; thus one man was paid 14s 'because the

King pulle forth his tooth', and another 18s for giving the King leave to bleed him.

It was no doubt James's interest in medicine that led him to investigate alchemy, in the study of which he was assisted by an Italian or Frenchman named John Damian. At Damian's suggestion, the king established and equipped an alchemical laboratory in Stirling Castle, and the accounts of the Lord High Treasurer of Scotland show that the venture cost a good deal of money. Though not inclined to be parsimonious, James quickly found that Damian and his alchemical experiments were a financial liability to be reckoned with, so, as Read puts it, he 'transmuted his alchemist into an abbot, in order to endow him with an emolument and provide him with the necessary leisure for his alchemical projects'. However, Damian seems to have found his salary inadequate, for he borrowed money from the king from time to time, and also received £210 for his expenses in connexion with a visit to the Continent on alchemical business. What may be even more significant were the very large amount of whisky the alchemist found it necessary to employ in his search for the philosophers' stone, and the occasional puncheon of wine for making the quintessence.

From the Treasurer's accounts for the years 1501 and 1508, Read has extracted the following figures, which show the cost of materials and apparatus used in Damian's researches:

	£	s	d
Alchemist's attire			
damask gown	15	16	0
velvet short hose	4	0	0
scarlet hose	1	10	0
Alembic, silver	6	19	9
Alum			7 per lb.
Aqua vitae			
small		6	0 per quart
ordinary		8	0 per quart
thrice-drawn		12	0 per quart
Bellows, small		1	0
Cauldrons, 18-gallon	1	1	0
Cinnabar		16	0 per lb.

	£	s	d
Flasks, large		4	0 and 6s 8d
Flasks, small		2	0
Glass, cakes of		5	0
Gold	6	10	0 per oz.
Litharge		5	0 per lb.
Mortar, metal, 53 lb	3	9	0
Mortar, brass, with pestle	1	1	0
Orpiment		6	0 per lb.
Pitchers, earthenware		1	3
Pot, large earthenware		1	0
Quicksilver		4	0 per lb.
Sal ammoniac	1	15	0 per lb.
Saltpetre			3 per oz.
Silver		14	0 per oz.
Sugar		1	6 per lb.
Sulphur		8	0 per stone
Tin		1	2 per lb.
Verdigris		6	0 per lb.
Vinegar			4 per gallon
White lead		2	0 per lb.

These figures are in contemporary Scots money and would have to be multiplied, by perhaps six, to give their modern equivalents.

Expenses for fuel and labour were also involved, for the furnaces were kept alight day and night on charcoal, wood, peat, and coal. The furnacemen received one shilling a day.

Though we do not read that Damian was ever successful in his operations, he remained on good terms with the king, and the two often used to dice and play cards together. Perhaps James liked the spirit of a man who could not only undertake laboratory experiments but embark upon an attempt at flying. One day Damian had a pair of wings made with feathers, strapped them upon himself, and 'took off from the lofty battlements of Stirling Castle for a flight to Paris'. His optimism was unjustified, for he fell to the ground and broke a leg; but he afterwards explained that some of the feathers used in the wings were those of barn-door fowls little accustomed to flight. Had eagle feathers been used exclusively he would no doubt have touched down at Le Bourget.

James IV was killed at Flodden in 1513, but the alchemical tradition was carried on by an adventurous Scot named Alexander Seton, whose story is related below (p. 223) in connexion with that of Michael Sendivogius (p. 281).

In an account of his researches into Scottish alchemy, Read has also published some curious facts about the interest taken in alchemy by John Napier of Merchiston (1550–1617), the inventor of logarithms. No doubt this interest arose from the fact that John's father, Sir Archibald Napier, had been master of the mint of James VI, but, however that may be, John himself conferred at Edinburgh with a German adept named Daniel Müller in 1607–8. Napier wrote an account of the conversations, from which it appears that he had already been a student of alchemy for some time, though he was not convinced that alchemical claims could be justified. Of the first interview he wrote:

> Upon Saterday the .7. of Nouember .1607. years, I Jhon Napeir Fier of Markeston, came to confer with Mr Daniel Muller, Doctor of Medecine, and student in Alchymie aneint our phylosophicall matters, not knowinge that he was sicke, and finding that he was diseased of the goute ... I remoued my compagnie, and sate done before his beddside, then he burst foorth in thise wordes. Sr you ar occupied in alchymie, I haue been this manie years ane verie earnest student therinto, and haue attained to the knowledge therof. I haue pressed to haue diuerted you from your wronge opinione, so farr as I durst be plaine, but now Sir I will be plaine, knowinge that you ar a man who fears God and will be secret, and that you will be good to my wyfe and bearins in case this diseases shall take me away.

Müller went on to relate that he had sent a friend, not yet returned, to Istria to bring a supply of Istrian cinnabar, because on one occasion when he broke a small piece of such ore

> the crude Mercurie flowed foorth without fyre, with this I perfited the phylosophicall worke as you may doe with the lyke, for this mercury beinge taken with fine silver which neuer did finde fyre and enclosed in ane matrix will become blacke within the space of 40 dayes, and thereafter will become white ... then you must joyne it with his ferment, to wite with fyne gould that neuer did finde the fyre ... and instantly that whit stoofe of mercury and luna will deuore up the gould, and at this coniunctione or fermentatione endeth the

first worke called *opus Lunae*, and beginneth immediately the second called *opus Solis*. In this *opere Solis* your worke becomes blacker than in *opere Lunae*, and then white, and at last reede. Both this workes ar performed in ane year. ... Now when I hard this thinges and had sayd unto him, My Lord, that matter is maruailous, if you be sure of the treuthe thereof by practice. He answered with earnestness, in treuthe I have practised it to the ende and made proictione, and found it true. Againe when I demanded him how it fortuned that he did not multiplie his stuffe and keeped the same, he answeared, I laiked [lacked] crud mercury without which it can not be multiplied againe.

This story has a familiar ring, so that it comes as a surprise that the trusty friend, whom we should expect never to have been heard of again, returned to England in the spring of 1608, bringing a great store of the Istrian cinnabar. Müller hastened to acquaint Napier with the joyful news, and when the friend arrived in Edinburgh a few days later presented him with a little of the ore, 'also ane uerie smal parte of Luna minerall vnfyned, but I purchased mor bothe of Scotes and German Luna. As for Sol minerall we have enough in Scotland.' There, unfortunately, the story comes to an abrupt end. Napier notes that he hopes to get time and opportunity to undertake the work, with the blessing of God, and to perform the same to His glory and to the comfort of His servants – but the rest is silence.

Read has, however, unearthed an alchemical letter written by the same Daniel Müller to the Earl of Argyll, in the form of a copy made by Patrick Ruthven (1584–1652), whose alchemical commonplace book is still preserved in the library of the university of Edinburgh. The first part of the letter gives directions for making a certain white powder, and Müller then says:

If you pleas heer to breake of your worke, then may you by verteu of this whitnes make daily new Moones [supplies of silver] at your pleasure, but better it wer to byde a letle longer, and then you shall see this whitnes turne into read and so by letle and letle, it will wear to a deep sanguine reed, in such graine as you cannot imagine a deeper, and this is called crocus solis, wherwith you may dye eurie Imperfect body into the Naturall colour of the Sonne and then is your wisshed worke at ane ende, and now thou mayest giue Vulcan leaue to sport him for a tyme, till thy farder occasions. If ye will trye whither thou hast wrought wysly, take one parte of thy reed powder and first poiect it upon . io . partes of thy reserued Sonn raies, and it will all become Meadsone [medicine] of Metals, and then proiect one parte of that vpon . io . partes of Mercurie, and tho shalt see thy Meadson will turnee this letle starr into a bright and shininge Sonne.

The 'Meadson' could also be relied upon to heal all manner of diseases, restore the sick to health, preserve the whole from sickness, and continue them both in an assured state of health until the hour appointed by God to call them hence. Perhaps it is unkind to recall that the Herr Doktor was a martyr to gout, but the alchemists were never noted for consistency. Of imagination they had plenty, of logic very little – as may be illustrated by their search for the alkahest. This name was introduced by Paracelsus for a spirit 'which acts very efficiently upon the liver', but a later alchemist, Van Helmont (1577–1644) described it as a universal solvent dissolving all bodies 'as warm water dissolves ice'. It is hardly credible that the alchemists continued to search for this marvellous substance even after Kunckel (d. about 1702) had pointed out that if the alkahest dissolved all bodies, it would dissolve the vessel into which it was put.

The great dissolvent of nature has been discussed often [he says]. Some say its name means *alkali est*, it is alkali; others say it is derived from the German *All-Geist*, universal spirit, or from *All ist*, it is all. But I believe that such a dissolvent does not exist, and I call it by its true name *Alles Lügen ist*, all that is a lie.

ALEXANDER SETON AND MICHAEL SENDIVOGIUS

Not long after Edward Kelly met his end, there rapidly crosses the alchemical scene an actor in the drama who, in his brief appearance, attracted most of the limelight by his spectacular adventures. He was a Scot named Alexander Seton, and may have come of a noble Scottish family. The circumstances of the two or three years of his life of which we have knowledge make him one of the most romantic characters of alchemy, in which there was never any lack of romance. The story starts appropriately enough on a stormy coast near Edinburgh, where in 1601 a Dutch ship was wrecked off the village of Seton. The pilot, Jacob Haussen, and the crew were rescued by the villagers, headed by Alexander Seton, of Seton House. Seton put Haussen up for a few days and showed him every kindness, finally arranging for his return to Holland and expressing the hope that they would meet again, a hope warmly reciprocated by the Dutchman.

In the following year, 1602, Alexander Seton made preparations for a continental tour and set off for Holland in the early spring, making first for the little town of Enkhuizen, twenty-eight miles from Amsterdam, where his Dutch friend was then living. Haussen received him with pleasure, and persuaded him to stay for several weeks, during which time the friendship became very cordial. So much so, that when at length Seton decided that he must continue his tour, he would not leave without presenting Haussen with a souvenir. This took the form of a piece of gold, which Seton prepared from a piece of lead of the same weight before the Dutchman's astonished eyes.

Such an exciting event could not be kept secret. Haussen told a local doctor, one Van der Linden, about the transmutation he had witnessed, and even gave him a small portion of the alchemical gold, which was inherited by the doctor's grandson, a historian of medicine, who relates this first part of the story. From Enkhuizen, Seton went to Amsterdam and Rotterdam, performing transmutations in each of those cities; then he took ship for Italy. His movements in Italy cannot now be traced, but late in the same year he travelled through Switzerland to Germany, in the company of Professor Wolfgang Dienheim of Fribourg. Dienheim describes his companion as a singularly witty man, not very tall and rather plump, possessed of a sanguine temperament, a high colour, and a brown beard cut in the French fashion. He brought with him only one servant, William Hamilton, who could be recognized anywhere by his red hair and beard. At Zürich, Dienheim and Seton hired a boat to take them to Basel, occupying the voyage by discussing the claims of the alchemists, in which Dienheim had no belief whatever.

Seton promised to reply to the professor's arguments not by further arguments but by an ocular demonstration of transmutation as soon as an opportunity occurred, saying that he hoped someone else would attend – someone whom he expected and who, like Dienheim, was a sceptic. This other person turned out to be Jacob Zwinger, member of a Swiss family noted for learning, and in his company they went to the house of a goldminer, carrying a few plates of lead, a crucible borrowed from a jeweller, and some ordinary sulphur which they bought on the way.

Seton ordered a fire to be lit, the lead to be placed in the crucible together with the sulphur, and the crucible to be covered with its lid. The lid was raised from time to time so that the mass within could be stirred. Seton himself touched none of the materials or apparatus, but chatted with the others while the heating was in progress. At the end of fifteen minutes he said, 'Throw this scrap of paper into the melted lead, well in the middle, and do not let any of it fall into the fire.' In the paper was a fairly heavy powder of a lemon-yellow colour, but so little of it that it could scarcely be distinguished. Dienheim says that though he and Zwinger were as incredulous as doubting Thomas they did what they were told. After further heating for a quarter of an hour, Seton ordered the miner to quench the crucible with water; which when he had done there was left a mass of gold equal in weight to the original lead. The jeweller attested its purity, which excelled that of Hungarian and Arabian gold, and Seton said mockingly, 'Now where are you with your pedantries? You can see the truth by the fact, which is more cogent than anything else, even your sophistries.' He then cut off bits of the gold and gave them to Dienheim and Zwinger.

Zwinger confirmed Dienheim's story in every detail, in a letter he wrote in 1606 to Emanuel Koenig, professor at Basel, who printed it in his *Ephemerides*. This letter also informs us that Seton made a second transmutation before leaving the town, this time for the goldsmith Bletz.

From Switzerland Seton went to Strasbourg, where it is probable that he was the anonymous alchemist who unintentionally brought grave misfortune to a blameless goldsmith named

Gustenhover. In the summer of 1603, a stranger arrived at Gustenhover's house, gave his name as Hirschborgen, and asked permission to carry out some work there. The goldsmith agreed, and when the stranger left some time later he made his host a present of a red powder and showed him how to use it to make gold. Vanity proved Gustenhover's undoing. He could not resist the temptation to boast of his good fortune and to demonstrate transmutation before several of his friends and neighbours, with the result that throughout Strasbourg a rumour ran, 'Gustenhover has discovered the secret of the alchemists! He has made gold!' The city council decided to investigate the matter and appointed three of its members to visit the goldsmith and make inquiries. Gustenhover was not in the least perturbed; he not only made transmutations under their eyes but invited them to carry out the experiment themselves, which they did, one after the other, with equal success.

By this time the news had reached the Emperor Rudolf II at Prague – the Emperor whose interest in alchemy had earned him the sobriquet of 'the German Hermes' – and Gustenhover was peremptorily summoned to the court. By this time his supply of the powder was exhausted, but his protestations that it had been given him by an unknown stranger and that he himself was ignorant of the method of preparing it were swept away by the avaricious monarch, who demanded that he should make gold at once. Gustenhover took the only course open to him, and fled; but he was pursued and captured, and spent the rest of his unhappy life immured in the White Tower of Rudolf's fortress.

Meanwhile Seton had moved on to Offenbach, near Frankfurt-am-Main, where he lodged under an assumed name with a merchant called Koch. In a letter to a friend, Koch says that before leaving Offenbach, Seton expressed a wish to teach him the art of alchemy. The usual apparatus and materials were prepared, but Seton would take no hand in the matter himself; Koch was to do everything. An ounce of mercury was placed in a crucible, and Koch added to it about three grains of a greyish-red powder given him by Seton. The crucible was then filled with potash to about half its depth and gently heated. More charcoal was placed in the furnace and the crucible was subjected to great heat for about half an hour. When it had become red-hot, Seton told Koch to add a small pellet of yellow wax, after which the crucible was quenched and broken. The residue proved to be a mass of gold weighing six ounces three grains. Koch had the gold assayed in his presence, and cupellation showed that it was of over twenty-three carats, the remainder being silver.

As usual, Seton seems to have disappeared immediately after the experiment, and it has been suggested that he regarded himself as a kind of missionary, dedicated to the task of converting those sceptical of alchemy to a state of belief by carrying out transmutations in their presence, or, better still, by getting the unbelievers to perform the work themselves under his instructions. In any case, his habit of passing on from city to city, never staying long in one place, attracted enough attention for him to be styled the Cosmopolite, under which name he became generally known.

From Frankfurt, Seton went to Cologne, where he commanded his servant Hamilton to try to get him into touch with local alchemists. The adepts did not seek publicity however, and the search was not at once successful. Then the resourceful Hamilton thought that distillers, who were in a sense practical alchemists, might provide the desired information, and from one of them he got the name of an amateur alchemist, named Anton Bordemann. Seton at once called on Bordemann, and arranged to lodge at his house. But Cologne was overrun with alchemical dabblers or 'puffers', who had made themselves a laughing-stock and were now keeping very quiet, and this state of affairs led Seton to be less prudent than usual. On 3 August 1603 he went into an apothecary's shop to buy some lapis lazuli, but the specimens in stock were not of good quality and the assistant advised him to call the next day, when a fresh supply was expected. There were several other customers in the shop, two of whom guessed that Seton wanted the lapis lazuli for alchemical purposes and told him that the citizens of Cologne had seen too many failures to retain any faith in alchemy. Seton rose to the bait, and hotly maintained that there did exist adepts competent to prove that, disbelieve as they may, the worthy citizens were wrong. At this there was a burst of laughter, and Seton strode from the shop furious with anger.

He was still fuming when he arrived at Bordemann's house, and made up his mind that those who had laughed at him should be made to hold their tongues. The following morning he returned to the apothecary's shop, purchased some of the better pieces of lapis lazuli that had now arrived, and then asked for some 'glass of antimony' (antimony oxide). Expressing doubt about the quality of the sample shown him, he said he would like to test it, so the apothecary told his son to conduct Seton to the workshop of the goldsmith Johann Lohndorf near the church of St Laurence. Lohndorf placed the glass of antimony in a red-hot crucible, and in the meantime Seton had drawn from his pocket a piece of paper containing a powder; this powder he separated into two portions with the tip of a knife, and gave one portion to the goldsmith with instructions to throw it on the melted glass of antimony. After a few moments the crucible was withdrawn from the fire, and in the bottom of it was found a button of gold. The incident was witnessed not only by the goldsmith who had carried out the work, but by the apothecary's son, two craftsmen working in the room, and a neighbour.

Herr Lohndorf was taken aback, but declared that he was not convinced. He suggested that the experiment should be repeated with the substitution of lead for the glass of antimony, at the same time furtively slipping into the crucible a morsel of zinc, which has the power of rendering gold brittle. Having, as he thought, thus ruined the experiment beforehand, Lohndorf sat back to watch the adept's discomfiture. His malice was, however, frustrated, for the button of gold was again perfectly pure, ductile, and malleable.

No one in the city of Cologne was more delighted at the result than Seton's host Bordemann. He had had his share of the raillery and ridicule aimed at the local alchemists, and, even though he had not been successful himself, he now had the honour of lodging the man who had sent the barbed witticisms winging in the opposite direction. The art had been rehabilitated, and he with it. It was no doubt at his instigation that Seton determined to try to convert a more doughty adversary of alchemy, one Meister Georg. Meister Georg was a surgeon and a learned man whose opinions and judgements were much respected, but his incredulity where alchemy was concerned amounted to prejudice. Seton therefore realized that it would be hopeless to secure an interview with Georg if he admitted his real intention, so invited the surgeon to meet him to discuss medicine and anatomy. Georg accepted the invitation, and, when they met, various medical topics engaged their conversation until Seton remarked that he knew how to cut away proud flesh, even down to the bone, without disturbing the nerves. The surgeon's professional interest was at once awakened, and he told Seton that he would very much like to see such an operation. This was Seton's chance, for which he had contrived, and he replied that nothing could be simpler. He sent Georg's assistant to procure some lead, some sulphur, and a crucible, but as bellows and a furnace were also necessary and were not available in the house where they were meeting, Seton proposed that they should go to the workshop of a neighbouring goldsmith. While waiting there for Georg's assistant, Seton got into conversation with men working in the room and asked them if they would like to learn how to convert iron into steel. They said they would be pleased if he would show them, so some old broken pliers were found and heated in a crucible. By this time the assistant had arrived with the other crucible, into which he had already put the lead and the sulphur; this crucible also was placed on the fire. The bellows were set going, and at the proper moment Seton produced his screw of paper containing a red powder, half of which he put into each crucible, at the same time ordering that charcoal should be placed above the crucibles and the heat of the fire increased. After a few moments, the lids of the crucibles were lifted, when the surgeon's assistant exclaimed with astonishment that his crucible contained gold. A similar cry of amazement came from the workmen, who found that instead of turning to steel their pliers had become gold.

The goldsmith's wife was called. She was an expert on the precious metals and their alloys, and had no doubt about the quality of this alchemical gold; in fact she offered to buy it for eight thalers after she had submitted it to a variety of tests. But the news had already leaked out and a crowd was gathering, so Seton and Georg slipped away.

Once they had got clear, Meister Georg said, 'So that was what you wanted to show me?' 'Certainly,' replied Seton. 'I learned from my host that you were a declared enemy of alchemy, and I wanted to convince you by unanswerable proof. I have done the same at Rotterdam, Amsterdam, Frankfurt, Strasbourg, and Basel.' 'But, my dear Sir,' said Georg, 'you are surely very imprudent to act so openly. Should your operations come to princely knowledge, your freedom will be in peril, and you will be held prisoner until you reveal your secret.' 'I am well aware of that,' admitted Seton, 'but Cologne is a free city. In any case, if a ruler were to seize my person, I would die a thousand deaths rather than reveal the secret to him.'

From a convinced sceptic, Meister Georg had now become a convinced believer. He was told by his friends that he had merely been the dupe of a clever charlatan, but all he would say was, 'What I have seen, I have seen.' Others less charitable accused him of having been bribed by the alchemist to bear false testimony. To this charge he did not deign to reply, and indeed the known probity of his life was sufficient refutation.

This striking conversion made, the Cosmopolite travelled to Hamburg, where he made a number of transmutations, and then on to Munich. If he left this city, too, secretly and swiftly, it was this time for a different reason; he had fallen in love with the beautiful daughter of a Munich worthy and eloped with her to Krossen towards the end of 1603. At that time, Krossen was the seat of the court of Christian II, Elector of Saxony, and hearing that Seton had arrived in the town, Christian sent for him. But Seton was on his honeymoon and made excuses, sending William Hamilton, equipped with some of the powder, as his deputy. Hamilton made a projection before the assembled court, with entirely satisfactory results; the gold withstood all tests. Perhaps Hamilton was frightened by his own success and foresaw what was likely to come from so cruel and avaricious a prince as Christian; or perhaps he felt that, now Seton was married, it would be a good moment to leave his service; but for whatever reason the two parted company. Hamilton returned to Scotland and is never heard of again.

If Hamilton had forebodings, they were well founded. Christian again invited Seton to the court and this time there could be no refusal. At first, the Elector appeared to be friendly, but it soon became clear that he would not be satisfied with demonstrations of transmutation: he demanded the secret of the powder. The Cosmopolite was steadfast in his refusal to part with it, so when sweet reasonableness failed, Christian had recourse to sterner measures. Seton was put on the rack, scourged, pierced with pointed irons, and burnt with molten lead; but all in vain. He would reveal nothing. Christian would doubtless have continued the torture still further, but Seton was now at death's door, and were he to die the last chance of wresting the secret from him would be lost. He was therefore put in solitary confinement in a gloomy dungeon constantly guarded by relays of soldiers.

It is at this point in the story that a new character comes on the scene, namely Michael Sendivogius or Sensophax, a gentleman of Moravia. A student of the Hermetic science, he was concerned at the news of Seton's imprisonment, and having influential friends at Christian's court was given permission to visit the prisoner; this he did on several occasions but always under the watchful eyes of the guards. They talked on alchemy, but Seton would say little. After some time, when the soldiers had grown accustomed to Sendivogius's visits, he and Seton were left to themselves, and Sendivogius immediately communicated to Seton a plan he had devised for the latter's escape. It seemed feasible, and Seton agreed that it should be tried. Sendivogius thereupon hurried to Cracow to raise funds by selling a house he possessed in that city, and quickly returned with a large sum of money. Again using his influence at court, he was given leave to live in the prison in which Seton was held, and set to work to gain the confidence of the soldiers. Judiciously bestowed tips had their effect, and on the day chosen for the attempt at escape he regaled the guards so lavishly that by the time night fell they were all dead drunk. Seton's tortures had rendered him too weak to walk, so Sendivogius picked him up and cautiously made his way out of the tower with his feeble burden. They were not challenged, but speed was vital, so after a hurried call at Seton's house to secure the remains of the precious powder, they entered a waiting post-chaise that Seton's wife had procured, and the trio drove off at full gallop to security across the Polish frontier.

They did not stop until they reached Cracow, where Sendivogius invited the Cosmopolite to redeem a promise made in the prison. Seton had said that if the rescue were successful, he would give his rescuer enough to provide for himself and his family for the whole of their lives. Sendivogius took this to mean that Seton would reveal the method of making the powder, but Seton refused to go so far; the promise had referred only to a gift of sufficient of the powder to fulfil the literal sense of the promise. With this, Sendivogius had to be content, and when Seton was approaching his end, only a short time after his rescue, he kept his word by handing over the whole remaining stock of the powder that had wrought not only transmutations but his own destruction.

The death of Seton occurred either late in December 1603 or on New Year's Day 1604, so that the events narrated were crowded into little more than two years. It is difficult to know what to make of those events. Rejecting as we must the hypothesis that Seton effected genuine transmutations, there would seem no alternative but that he was an enormously adroit and plausible imposter who chose this singular but very successful method of winning popular admiration. That, at least, is the opinion of Figuier, the nineteenth-century French historian of alchemy, who adds that history provides other examples of men ready to sacrifice their wealth, their talents, and even their lives in propagating errors, merely to obtain celebrity.

Seton's rescuer, Michael Sendivogius, was born in Moravia in 1566, but the fact that, as we have seen, he possessed a house in Cracow has sometimes caused him to be described as a Pole. This house he inherited from one Jacob Sandimir, whose natural son he is said to have been. As a youth he showed considerable industry and quickly became an expert on mining and the manufacture of pigments, but he does not appear to have claimed any personal alchemical skill before his acquaintance with Seton. This was in the thirty-seventh year of his age, and for some time previously he had been living in style at Dresden. It seems likely that his lavish expenditure there was too great for his resources, and that his interest in Seton was based on the hope that, by rescuing the captive, he might succeed in winning from him the secret that Christian had failed to extract. The money he received from the sale of the Cracow house had now been spent, and Sendivogius was beginning to feel very anxious about the future. In the renewed hope of gaining the secret of the powder, he married Seton's widow, but he found that she knew no more than he; she did, however, give him an alchemical manuscript that Seton had written. On reading this work, Sendivogius thought he had found a means, not of preparing the powder outright, but of multiplying the stock he already possessed; unfortunately for him the attempt was not successful and he wasted much of the powder in the experiment.

He nevertheless would not abandon his ostentatious mode of life and performed several transmutations in public to establish his reputation as an adept. When travelling, he carried the powder in a small golden box, always being careful to hand the box personally to his major-domo, who wore it under his clothes from a gold chain round his neck. In point of fact, this was mere display, the bulk of the powder being hidden in a secret recess in the running-board of his carriage.

By this time, the story of Sendivogius's alchemical successes had become widely known, and he was invited to court by the ever optimistic Rudolf II. On arrival he took the precaution of saying at once that he had no knowledge of the method of preparation of the powder, and for once the Emperor was satisfied with such a disclaimer. Sendivogius demonstrated the method of working with some of the powder he had obtained from Seton, and then suggested that Rudolf should perform the experiment himself. It was fully successful, and Sendivogius was given the title of counsellor to his Majesty.

With commendable restraint, Sendivogius sought no greater favours at Prague, but set off for Poland in response to an invitation from King Sigismond. On his passage through Moravia, however, he was ambushed by a local magnate and imprisoned, being told that the price of his release was the secret of the philosophers' stone. Having not the least desire to suffer the same fate as Seton, he resolved to escape, and from some source or other managed to procure a file with which he filed through the bars of his cell; but the height was too considerable for him to jump to the ground, so he tore his clothes into strips and made a rope by which he lowered himself sufficiently to drop the remainder of the way to safety. Running through the countryside, he at last found a good Samaritan who gave him shelter, and he then lost no time in sending a message to the Emperor denouncing the villain who had waylaid him. The Emperor was indignant that one of his counsellors should have been so maltreated, and the offender was made to pay Sendivogius a large sum in damages, and also to give him a country estate.

After this disturbing occurrence, from which as a matter of fact Sendivogius derived very timely assistance in his affairs, he continued his journey to Poland and arrived safely at Warsaw. Here he made a number of transmutations before Sigismond, but his supply of powder was now getting very low and he welcomed an invitation to the court of Duke Frederick of Würtemberg. Frederick received him with great friendliness, and was so astonished by two transmutations that he created Sendivogius Count of Nedlingen. Appropriating the name Cosmopolite from Seton, the new count returned to his extravagant way of life, and his reputation as a successful alchemist grew greater every day. Everything seemed set fair, but, unknown to himself, Sendivogius had made a bitter enemy. This was a barber-turned-alchemist named Müllenfels, who, not being without ingenuity, had been engaged by Frederick as court representative of the art. He now found himself completely outshone by the new arrival, and plotted to overthrow the usurper and at the same time to relieve him of the powder.

Repressing all signs of jealousy, he missed no opportunity of expressing his admiration of the virtuosity shown by Sendivogius, and by fulsome praise, readily lapped up by vanity, he was not long in winning the confidence of the rival whom he hated. When he judged the ground well prepared, Müllenfels took Sendivogius aside, and in grave and confidential manner told him that Frederick was about to force him to reveal his secret; moved by avarice, the tyrant would refrain from no severity, no torture, until he had won the precious information that he coveted. Sendivogius was highly alarmed: he knew only too well the kind of treatment with which Müllenfels said he was threatened. After a moment's thought he decided to make for the frontier immediately, a decision warmly approved by Müllenfels, who gave him directions on the best route to follow. Hardly had he departed, in the small hours of the morning, when Müllenfels gave chase, accompanied by a dozen armed men on horseback. They soon overtook the fugitive, ordered him to stop, and robbed him of the powder and of a diamond necklace and other valuables, after which they allowed him to proceed. He was apparently too frightened to return to Frederick to complain, and went into hiding for over a year; but Müllenfels gained little from his treachery. News of the attack spread, and it was generally believed that the whole affair had been staged by the Duke of Württemberg himself. Nothing happened, however, until at length Sendivogius thought it safe to appear once more; he then went to Rudolf and demanded justice, explaining the plot that had been put into action against him.

Rudolf acted with vigour, and sent an express envoy to Frederick demanding that Müllenfels, whom Frederick had now restored to his former eminence, should be committed to him for punishment. Frederick felt or simulated great anger at the implied charge against himself, returned to Rudolf the valuables stolen from Sendivogius – except the powder, which he said he knew nothing about – and told the Emperor that he would deal with Müllenfels. He had a double grievance: Müllenfels had lost Sendivogius for him, and had also embroiled him with the Emperor. The alchemist therefore had short shrift and was hanged as soon as a gallows could be erected.

This series of events happened in 1607, and nothing further is heard of Sendivogius until 1625, when he reappeared in Warsaw. Müllenfels's theft of practically all his powder had put an end to his successful alchemical career, and he was now reduced to the tricks of the vulgar charlatan, selling marvellous nostrums, counterfeiting silver, and borrowing money from the credulous on the strength of false promises to make the philosophers' stone. A pitiable figure, he lingered on till 1646, when he died at Cracow aged eighty.

The fact that Sendivogius adopted Seton's nickname of Cosmopolite has caused some confusion of the two, a confusion increased by the further circumstance that Sendivogius published the manuscript work given him by Seton's widow under the pseudonym *Divi leschi genus amo*, which cryptographers soon perceived to be an anagram of Michael Sendivogius. A little later he also published a 'Treatise on Sulphur', of which he is believed to have been the true author, under another anagrammatic pseudonym, namely *Angelus doce mihi jus*. After the latter book had been published, Sendivogius noticed that there were rather glaring discrepancies between it and the former one; he therefore altered the text of the 'Treatise on Sulphur' for a second edition, to bring the two works into line.

The part played by Rudolf II in the careers of Seton and Sendivogius, no less than in the lives of other alchemists of the time (pp. 207, 237), may make us curious to know something of this gullible monarch, who was born in 1552 and was Holy Roman Emperor from 1576 until his death in 1612. We are told that he was of a mild and reserved disposition, not fond of bodily exercise but showing a genuine interest in all branches of art and science, and a keen collector of pictures, statuary, jewellery, and curiosities of all kinds.

> But he was no mere collector; there was hardly an art or a science of which some distinguished representative was not to be found at his Court; he was a great reader of Latin verse, and a friend of historical composition; and he entered with special interest into mathematical, physical, and medical studies. Chemistry and astronomy – with their then inseparable perversions, alchemy and astrology – irresistibly attracted his speculative mind; and ... Tycho Brahe and Kepler enjoyed his patronage.

At first fulfilling his duties as head of the state, he gradually allowed his private tastes to distract him from the business of government, and by about 1597 or 1598 he was showing unmistakable signs of melancholia and madness; he would now not even trouble to sign important papers. Things rapidly grew worse, until instead of exerting any rule, the Emperor was himself being ruled by his valets; even ambassadors could not obtain audience, and the closing years of the reign were marked by acts of cruelty that could have been inspired only by the suspicions and craftiness of insanity.

When Rudolf had given up effective government he shut himself in his castle at Prague and devoted more and more of his time to alchemy. The servants became laboratory assistants, and the court poet was charged with the composition of laudatory odes on successful adepts and with the versification of alchemical treatises. All alchemists, of whatever race or station, were sure of a welcome. They were first interrogated by Thaddeus von Hayec, the Emperor's physician and himself an adept, and if they were found to have sufficient knowledge of their subject they were given audience. Those who were able to show the Emperor some striking experiment were invariably well rewarded. But Rudolf did not depend solely on these chance visitors; as we have seen, he would invite or summon to his court any practitioner of the art whose fame reached him. The invitations were not always accepted. One alchemist, of Franche-Comté, refused to budge, saying with simple but unanswerable logic, 'If I know the secret of alchemy, I have no need of the Emperor; if I do not know it, he has no need of me.' Rudolf took this refusal in good part, and entered into a correspondence with the reluctant adept.

Rudolf inevitably became credited with possession of the philosophers' stone, an opinion that was felt to be justified after his death by the discovery in his laboratory of 84 cwt of gold and 60 cwt of silver, all neatly cast in the form of small bricks. Adjoining this treasure was a quantity of grey powder, which nobody doubted to be the Stone. Taking advantage of a favourable moment, the valet Rutzke made off with this powder, which he bequeathed to his family; but when trial was made of it only failure resulted.

Other royal patrons of alchemy of about the same time included Anne of Denmark, wife of the Elector of Saxony. In her castle she had her own private laboratory, which the German alchemist Kunckel – one of the earliest workers on phosphorus – described as the largest and finest he had ever seen. Marie de Médicis, queen consort of Henry IV of France, was sufficiently a believer in alchemy to give a prisoner in the Bastille, one Guy de Crusembourg, 20,000 crowns to prepare the philosophers' stone. No doubt she thought the money could not there be spent except in the way specified, but Guy thought otherwise. He escaped from the Bastille, taking the money with him, and the Queen was never able to discover either.

II

TWO FRENCH ALCHEMISTS: FLAMEL AND ZACHAIRE

NICOLAS FLAMEL

SINCE the Chinese, in spite of their diligent search, were apparently unsuccessful in preparing the pill of immortality, they might have done well to take lessons from a fourteenth-century Parisian alchemist, Nicholas Flamel; for he and his wife were reported to be alive and well in India in the seventeenth century, while in 1761 they attended the opera in Paris! Flamel was indeed the alchemist *par excellence* of France, and the story of his life does not need, though it explains, such later embellishments.

It is very probable, but not quite certain, that Flamel was born at Pontoise, eighteen miles north of Paris, about the year 1330. His parents were in only a modest station of life, but they were able to give him a sufficiently good education to enable him while still young to set up in business in Paris as a public scribe or notary. At first, he was to be found at the Charnier (ossuary) des Innocents among others of his profession, and when the guild of scribes moved as a body to the neighbourhood of the church of Saint-Jacques-la-Boucherie, Flamel went with the rest. His wooden booth, where he worked, was only two feet by two and a half, but he also had a house close by, called 'At the Sign of the Fleur de Lys', where pupils took lessons and where his calligraphers, assistants, and apprentices produced illuminated and other psalters, books of hours, and manuscripts of all kinds. Printing had not yet been invented, and Flamel thus combined the business of scribe with that of bookseller and publisher. His reputation stood high; he was even patronized by ladies and gentlemen of the court who wished to be taught how to sign their names. Illiteracy was perhaps more general in those days than now.

Soon after he had moved to the rue de Saint-Jacques-la-

Boucherie, or rue des Écrivains as it was called, Flamel was engaged by a Madame Lethas to copy a deed. She had been twice widowed, led a lonely life, and allowed Flamel to see that she would like to marry again. As the attraction proved to be mutual the wedding was not long delayed, and so Flamel won his faithful wife and companion Pernelle, of whom he always speaks with affection. She was over forty when he married her, but still good-looking, and was comfortably off. The couple led a life of great simplicity, dressing quietly, eating from earthenware vessels, and diligently performing their religious duties. They had a cook and housemaid, for some of the writers and apprentices fed in the house and there was a good deal of cooking to be done; but the domestics appear to have been well treated since they were still in Flamel's service at the time of his death, when he left them adequately provided for. Such details are necessary to the narrative, inasmuch as they help to show that the Flamels were thrifty and industrious people owning a prosperous business: it is thus not surprising that they grew rich, but their wealth was partly responsible for Flamel's later fame as an alchemist.

Though Flamel must often have had alchemical manuscripts to copy, he took no particular interest in the subject until one night, when he was in a profound sleep, an angel appeared to him in a dream, holding out before him a magnificent book of venerable antiquity. 'Flamel,' said the angel, 'look at this book. You will not in the least understand it, neither will anyone else; but a day will come when you will see in it something that no one else will see.' Flamel stretched out his hand to take the book, but both book and angel disappeared into a golden cloud.

Flamel took little notice of the dream and would probably have forgotten it altogether but for a remarkable event that occurred some time later. On a certain day in 1357 he bought from an unknown vendor an old book which he at once recognized as being the very book he had seen in his dream:

There fell into my hands, for the sum of two florins, a gilded book, very old and large; it was not of paper or parchment, as other books are, but made only of thin bark (as it seemed to me) of tender shrubs. Its cover was of copper, very delicate, and engraved all over with strange letters or figures. I could not read them but I thought that they might be in Greek or some other ancient language. The leaves of bark inside were covered with beautiful and very clear Latin letters, which had been inscribed with a steel point and coloured. The book contained three times seven leaves, for so they were numbered at the top of the leaves, the seventh leaf always without writing on it, but instead, on the first seventh leaf, had been painted a rod, with two serpents swallowing one another; on the second seventh, a cross on which a serpent was crucified; and on the last seventh were painted deserts, in the midst of which ran beautiful fountains, from whence there issued many serpents which ran hither and thither. Upon the first of the leaves there was written, in large capital letters of gold: ABRAHAM THE JEW, PRINCE, PRIEST, LEVITE, ASTROLOGER, AND PHILOSOPHER, TO THE NATION OF THE JEWS, BY THE WRATH OF GOD DISPERSED AMONG THE GAULS, SENDETH SALUTATION. After this it was filled with great execrations and curses (with this word MARANATHA, which was often repeated there) against every person that should cast his eyes upon it, unless he were Sacrificer or Scribe.

Flamel then describes how, feeling exempted from the curses by his profession of scribe, he examined the book further, but could gather nothing beyond the fact that in it the author was revealing to the Jews the art of transmuting metals into gold, so that they might pay the tribute due to the Roman emperors. The difficulty lay in the fourth and fifth leaves, which were filled with beautiful illuminated figures unaccompanied by any written explanation. Flamel guessed that they contained the secret of making the philosophers' stone, but that they could not be understood without a deep knowledge of Hermetic books and of the Jewish Cabbala.

The first figure of the fourth leaf represented a young man with wings at his ankles, holding in his hand a caduceus round which writhed two serpents, and with which he struck a helmet that covered his head; this young man resembled the pagan Mercury. Against him there came running and flying with outspread wings a great old man with an hour-glass on his head, and in his hands a scythe, like Death: terrible and furious, he would have cut off the Mercury's feet. A second figure on the same leaf showed a mountain on the summit of which was a fair flowering bush shaken by the north wind; its base was blue, its flowers white and red, and its leaves shining like gold. Around it, the dragons and the griffons of the north made their nests and dwellings.

On the fifth leaf was depicted a fair garden, in the middle of which a rose-tree in bloom climbed up a hollow oak. At the foot of the oak there boiled a fountain of most white water, which rushed down over the edge of an abyss: but before thus disappearing its waves had passed through the hands of many people, who dug the earth to seek it but who, being blind, did not recognize it, except those among them who considered its weight. On the other side of the fifth leaf was a king armed with a falchion, who caused to be killed in his presence, by soldiers, a multitude of little children, whose mothers wept at the feet of the pitiless slayers. The blood of these infants was gathered up by other soldiers and put in a great vessel, wherein the Sun and Moon came to bathe themselves.

Most of this symbolism is of the usual alchemical type: the figure of Mercury represents philosophers' mercury, and the old man with the scythe attempting to cut off Mercury's feet is Saturn or lead, which, when added to mercury, renders it 'fixed' or solid. The white and red flowers are symbolic of the White and Red Elixirs, and the fountain of most white water is again philosophers' mercury. The blood of the innocents represents the mineral spirit of metals; and so on.

Flamel spent days and nights in the study of this precious book, hiding it from all eyes even though he could not comprehend it. Pernelle, however, soon perceived that her husband had something on his mind, for he was sad and she often heard him sigh. She was gently persistent in questioning him, and at last he confided in her. Pernelle faithfully kept the secret, though she could do nothing to help to decipher the mysterious instructions. The exasperating fact was that, though the first few pages were intelligible, Flamel could not begin to work because the nature of the starting-material was not given there, but was presumably hidden in those parts of the book that neither he nor Pernelle could understand.

At length, Flamel came to the conclusion that he would never succeed unaided, and resolved to seek the advice of those better versed in the language of alchemy. He proceeded very discreetly, not exhibiting the book itself but making careful copies of some of the figures in it and putting them on view in his shop. Here they attracted much attention, but no one could interpret them, and when Flamel said he thought they contained the secret of the philosophers' stone he had to suffer a good deal of banter. However, among visitors to the shop was one Anselm, a licentiate in medicine who much fancied himself as an alchemist. He was familiar with Flamel's style, and guessed that the drawings had been made by the notary from originals not on view; but when taxed with this, Flamel would give only evasive replies. Anselm therefore had to be content with explaining the copies, and as he had alchemical patter on the tip of his tongue his explanations were glib and convincing. Acting upon them, Flamel began a series of unsuccessful experiments that occupied his spare time for twenty-one years. He was getting very discouraged when suddenly a happy inspiration came: the book was written by Abraham the Jew for men of the Jewish race. Probably, therefore, no one but a Jew would be able to understand it.

Flamel at once decided to make a pilgrimage to Saint James of Compostella, near Corunna in Spain, hoping by this pious act to receive the favour of discovering in a Spanish synagogue some learned Jew who could give him the true interpretation of the symbolic figures. Pernelle willingly agreed to the suggestion, and in 1378 he donned a palmer's gown, took staff and cockleshell, and set off on his long journey. This was accomplished without incident, and, his vow fulfilled, Flamel began to frequent the synagogues as he had planned. For a long time – nearly a year – he had no success, and at length decided that he must return to France. He had reached León on his way home, when in that city he made the acquaintance of a fellow-countryman, a merchant of Boulogne. Flamel told the merchant the object of his journey, and that he had not been able to find a Jew who could help him, whereupon the merchant offered to introduce him to a friend of his, a converted Jew named Maître Canches, who was a learned cabbalist. Flamel felt that this was an answer to prayer, and a meeting was arranged; the merchant had to act as interpreter, for Flamel could not speak Spanish and Maître Canches could not speak French. But as the story unfolded, Maître Canches grew more and more interested, until his impatience at the delay caused by interpreting made him suggest to Flamel that they should continue the conversation in Latin, which they did. At the sight of Flamel's drawings his excitement knew no bounds, for he identified them as coming from the *Asch Mezareph* of the Rabbi Abraham, a book that the cabbalists had thought lost for ever. He at once began deciphering them, in a way quite different from Anselm's, and Flamel realized that, after so many years of disappointment, he was now on the right track. He and Maître Canches were firm friends from that moment, and when Flamel revealed that he possessed the original book from which the drawings had been copied, Maître Canches was all agog to return with him to Paris as quickly as possible.

To this end, they made for Oviedo and then for Sanson, where they embarked on a ship to save much of the long journey overland. The voyage was pleasant enough for Flamel, but Maître Canches proved to be a bad sailor and was very seasick. They went up the Loire as far as Orleans and landed there, but unhappily Maître Canches's seasickness had aroused some more serious form of illness; the vomiting grew more violent, and at the end of a week he died. Flamel says that he was *fort affligé* at the death of his friend, whom he buried in Holy Cross Church at Orleans, and then took the road for Paris.

In spite of his grief, Flamel could not help feeling at the same time a great satisfaction, for before his fatal illness Maître Canches had explained enough of the secrets of the book to enable Flamel to start on his experiments again with every prospect of success. Sunburnt from his journey, he was welcomed with joy by the faithful Pernelle, whose joy was heightened when she heard that his pilgrimage had not been in vain. 'He that would see the state of my arrival', wrote Flamel, 'and the joy of Pernelle, let him look upon us two on the door of the chapel of Saint-Jacques-la-Boucherie, close by the side of my house, where we are painted, myself giving thanks at the feet of Saint James of Compostella, and Pernelle at those of Saint John, whom she had so often invoked.' In this passage Flamel refers to a sculptured and painted portal erected at his expense in 1389; it was an expression of gratitude both for his safe return and for the happy outcome of the journey.

Although Flamel now had most of the necessary information, it took him three years of unremitting labour to achieve final success. The penultimate stage was reached: all that was left to be done was to heat the product in a glass flask or 'philosophical egg' set in an athanor. With beating heart, Flamel watched for the revealing colours. They came, and in the correct sequence: from grey to black, 'the crow's head', then from black to white, the white first appearing like a halo round the edge of the black, and the halo then shooting out white filaments towards the centre, until the whole mass was of a perfect white. This was the white elixir, and Flamel could wait no longer; he opened the flask, took out the elixir, called Pernelle, and prepared to make the trial. This was on Monday, 17 January 1382. Taking about half a pound of lead, he melted it in a crucible, and to it added a little of the white elixir, whereupon the lead was at once converted into silver, purer than the silver of mines.

Sure at last that he had achieved mastery of the Art, he replaced the rest of the elixir in the flask and continued the heating. Now the rest of the colours appeared one after the other: the white turned to the iridescence of the peacock's tail, this to yellow, the yellow to orange, the orange to purple, and finally the purple to red – the red of the Great Elixir.

And then [he says], following my book word for word, I made projection of the Red Stone upon half a pound of mercury, in the presence of Pernelle only, in the same house, the five-and-twentieth day of April following, the same year, about five o'clock in the evening; which I transmuted truly into about the same quantity of pure gold, most certainly better than ordinary gold, being more soft and more pliable. ... Pernelle understood it as well as I, because she helped me in my operations, and without doubt if she had undertaken to do it herself she would have attained to the end and perfection thereof. I had indeed enough when I had once done it, but I found exceeding great pleasure and delight in seeing and contemplating the admirable works of Nature.

Besides these two, Flamel made only three more projections, as a result of which he was believed to have made enough gold to account for the numerous acts of charity he performed. These included, in Paris, the foundation and endowment of fourteen hospitals, the building of three chapels, rich gifts to seven churches, and the repair of church buildings, while he made similar benefactions at Boulogne, probably the town where Pernelle was born. In his will, Flamel also left a legacy to the church of Notre-Dame at Pontoise. The extent of these gifts, though large, was not in fact beyond the limit of the fortune that a prosperous notary, publisher, and bookseller might have acquired, especially one of a frugal and industrious life; but popular opinion was unanimous in regarding Flamel as a triumphant alchemist who, an exception to the rule, used his mastery of the Art, not for his own advancement, but for the benefit of the poor and to the greater glory of God.

Flamel is said to have been strongly built and a little above the average in height. He had slender hands and his head was rather small, with a high forehead and large eyes deeply sunk. He had a straight nose, and an attractive mouth more ready to smile than to laugh, with lines at the side indicative of some asperity. The general impression was one of kindness and refinement. Pernelle was slim, and less tall than her husband; she had an oval countenance with fine and regular features. She died on 11 September 1397, having made her will a few days previously and leaving precise instructions about her funeral: the candles were to weigh altogether thirty-two pounds, the refreshments to cost £4 16s, and the sum of £8 17s to be expended on an anniversary mass. A further clause ordered that a pilgrim should proceed on foot to Boulogne-sur-Mer, there to have sung and said two masses in the church of Notre-Dame, making an offering in each case of a candle weighing twelve pounds. For carrying out this task he was to be paid £4.

The death of Pernelle deeply affected Flamel, and for a time he was inconsolable; but owing to family disputes over her will he had little time to brood, and by the time they were settled his grief had changed to a grateful and vivid memory of the long and happy years they had had together. He lived for another twenty-one years, not marrying again but devoting his days to good works and to writing on alchemy. Now old and growing feeble, he made his will on 22 November 1416, and engaged a mason to prepare a tombstone with an inscription composed by the alchemist himself. He died on 22 March 1417 and was buried in the church of Saint-Jacques-la-Boucherie towards the extremity of the nave. The church was secularized in 1790, during the Revolution, and was at first let by the revolutionary government to an industrialist; seven years later it was sold to a builder for 411,200 francs, part of the agreement being that he should demolish it. Fortunately, however, by an oversight nothing was said in the agreement about demolishing the tower, which the builder therefore left alone. For some time it was used as a shot-tower, but in 1836 it was purchased by the City of Paris, which its fine Gothic architecture still adorns (plate 23).

Demolition of the church led to the disappearance of Flamel's tombstone, along with many other relics, but by a quirk of fate it came to light many years later, when it was discovered in the shop of a fruiterer and herbalist in the rue des Arcis; the shopkeeper found its polished surface most convenient for chopping up his dried herbs. M. Guérard, who kept a pet shop, rescued it from this base office, hoping to sell it at a good price, but he could find no customer so handed it over to an antique dealer, M. Signol. It remained unsold for six years, and in the end M. Signol was pleased to part with it, for the trifling sum of 120 francs, to the Musée de Cluny, where it is still preserved (plate 24).

The stone is of marble, 58 centimetres long by 45 broad, and 4 centimetres thick. The upper portion shows carved representations of Christ, making the gesture of benediction with his right hand and holding in his left a globe surmounted by a cross; and of St Peter and St Paul, one on each side. Between the central figure and St Peter is the symbol of gold or the Sun, and the symbol of silver or the Moon occupies a corresponding position on the other side. The inscription in the middle of the stone reads:

Feu Nicolas Flamel jadis escrivain a laissie par son testament a leuvre de ceste eglise certaines rentes et maisons quil avoit acquestees et achetees a son vivant pour faire certain service divin et distribucions dargent chascun an par aumosne touchans les quinze vins lostel dieu et autres eglises et hospitaux de paris Soit prie po les trespasses.

Immediately above the wasted corpse are the words *Domine deus in tua misericordia speravi*, while the inscription below reads *De terre suis venus et en terre retourne Lame rens a toy Ihu qui les pechiez pardonne.* The *Ihu* is an abbreviation of *Jesus.*

After Flamel's death many people imagined that he must have left some of the philosophers' stone hidden either in his house in the rue des Écrivains or in one of the other houses he owned in Paris. His property was thus visited and ransacked by unauthorized persons and much damage was done, especially to the house in the rue des Écrivains, of which soon only the cellars were left. Even as late as 1560 hope had not been abandoned, for in that year the magistrate of the Châtelet district took possession, in the name of the king, of all property formerly belonging to Flamel, in order that a systematic and thorough search might be made. Greed and avarice overlooked two lines on the façade of 'At the Sign of the Fleur de Lys':

Chacun soit content de ses biens.
Qui n'a suffisance, il n'a rien.

It may have been Flamel's fame that led the town council of Dijon in 1455 to take strong measures on behalf of an alchemist named Pierre d'Estaing against a local landowner, one Jean de Bauffrement. D'Estaing, who was a physician as well as an alchemist, and a man of some standing, had been engaged by de Bauffrement to make gold alchemically and had been paid a large sum of money to defray the cost of the experiments. No success was forthcoming, however, and since he saw that his patron was becoming very restive d'Estaing thought it prudent to make an unobtrusive departure. He found refuge with the Dominicans at Dijon, but de Bauffrement discovered that he was there, entered the cloister by force, and carried him off.

This act of violence aroused considerable resentment among the citizens of Dijon, and the councillors laid the matter before the Duke of Burgundy. The outcome was that de Bauffrement had to stand trial and was heavily fined, but what became of d'Estaing is not known. The fact that the Duke of Burgundy and the town of Dijon took arms in support of an alchemist is, however, sufficiently remarkable to warrant the supposition that the status of alchemy in France may have been considerably raised by the widely current stories of Nicolas Flamel.

DENIS ZACHAIRE

AN amusing autobiography has been left to us by an alchemist passing under the name of Denis Zachaire, though whether he was really so called or wrote under a pseudonym is not certain. In any case his identity is unimportant, for the autobiography contains no new alchemical facts; its value lies in the picture it draws of the way in which alchemical research was carried out in the second quarter of the sixteenth century. It was several times printed between 1583 and 1740, in Latin, French, and German, but not in English. However, an English translation of a French edition of 1612 was made and published by Tenney L. Davis in 1926 and is the basis of the following abridgement.

Zachaire describes himself as a gentleman of Guyenne (Guienne, south-west France), and says that when he was about twenty years old he was sent by his parents to Bordeaux to attend the lectures on arts at the college there. He made such good progress that, although his father and mother had died in the meantime, his friends and relatives arranged for him to proceed to Toulouse to read law. But while at Bordeaux he had become interested in alchemy and had compiled from various sources a thick book of recipes or instructions for effecting transmutation. He itched to make experiments, and says:

> Immediately that I was arrived at Toulouse, I set myself to building small furnaces, being confirmed in all things by my master [a private tutor whom he had had at Bordeaux and who went with him to Toulouse]. Then from small I went to large, and soon I had a room entirely equipped with them, some for distilling, others for

subliming, others for calcining, others for dissolving in the water-bath, others for melting, in such sort that for a beginning I spent in one year the two hundred crowns which had been supplied to us for our support during two years of study. This was spent in setting up the furnaces and in buying charcoal, divers and infinite drugs, divers vessels of glass which I bought for six crowns a time, without counting two ounces of gold which were lost in practising one of the recipes, and two or three marks of silver for another, in which, if any of it was recovered, which was very little, it was so crude and so blackened by the force of the mixtures which the recipes had ordered to be added that it was almost entirely useless.

So, at the end of the year my two hundred crowns were gone up in smoke. And my master died of a continued fever which seized him during the summer by reason of the soot that he breathed and swallowed, for he was so desirous of accomplishing something worth while that he hardly ever left the room, where he made scarcely less soot than there is in the Arsenal of Venice where cannon are cast. His death was a great grief to me.

Having exhausted his money, he went to the paternal home, which was now his property and which was being looked after by caretakers, sacked them, and let the house for three years for 400 crowns. He then returned to Toulouse, in order to try out a recipe supplied to him by an Italian, who assured him that he had seen it tested.

I kept this man with me to see the outcome of his recipe, to practise which I was obliged to purchase two ounces of gold and a mark of silver. When these were melted together, we dissolved them in *aqua fortis*, then we calcined them by evaporation. We tried to dissolve them with divers other waters by divers distillations so many times that two months passed before our powder was ready to make the reprojection of it. We used as much of it as the recipe required, but it was in vain. . . . From all the gold and silver which I had used I recovered only half a mark, without counting the other costs, which were not small. So my 400 crowns were reduced to 230, and of this I supplied my Italian with twenty to go to find the author of the recipe, who he said was at Milan, in order that he might write back to us. After this I was at Toulouse all the winter awaiting his return, and I should be there yet if I had decided to wait for him – for I have not seen him since.

Zachaire had to leave Toulouse for the following summer, on account of an outbreak of plague, but at Cahors, where he stayed for six months, he obtained advice as to which of his recipes seemed the most likely to succeed. On his returned to Toulouse he tested them, but received only the same sort of profit as before, namely that his 400 crowns became 'augmented', as he wryly puts it, to 170.

But not for that did I cease always to pursue my enterprise. And the better to be able to continue it, I joined with an Abbé near Toulouse who said he had the duplicate of a recipe for making our great work which a friend of his who followed the Cardinal of Armagnac had sent to him from Rome and which he took to be perfectly assured. For this I furnished 100 crowns and he the like; and we began to set up new furnaces in a wholly different fashion to work on it.

They bought enough charcoal at once to keep the furnaces going for a whole year, but the powder they obtained at the end of that time proved entirely without effect when projected upon mercury.

I leave you to judge whether we were angry about it, especially Monsieur l'Abbé, who had already published to all his monks like a very good public secretary that it remained only to have a beautiful lead fountain melted which they had in their cloister to convert it into gold immediately that our result should be achieved. But he had it melted at another time to provide material in vain for a certain German who stopped at his abbey while I was in Paris.

In spite of all this he did not cease to wish to continue his undertaking, and he advised me to set about to get together three or four hundred crowns, and he would furnish the same amount, in order that I might go and live in Paris, a city which is today more frequented by divers operators in this science than any other in all Europe, and become acquainted with all sorts of people, to work with those in whom I recognized something worth while, and to divide it between us like two brothers.

Thereupon Zachaire found tenants for his whole estate, much to the anger of his relatives, and returned to Paris with 800 crowns in his pocket. The city was swarming with alchemists, and he soon made the acquaintance of more than a hundred of them. Hardly a day passed on which they did not assemble at one of their number's lodging, or in Notre-Dame, to discuss results obtained the day before.

Some would say, 'If we had the means to start again, we should do something worth while', and others, 'If our vessels had held we should be there', and others, 'If we had had our copper vessel perfectly round and well closed, we should have fixed Mercury with the Moon', and so on, for there was not one among them who had had any success, and not one who was not ready with an excuse.

However, Zachaire was impressed by one man who promised to show him how to extract silver from cinnabar.

Since he needed filings of fine silver we bought three marks of it and made them into filings. He mixed these with the pulverized cinnabar and made little pegs out of the mixture with an artificial paste, and heated them in a well covered earthen vessel for a certain time. When they were thoroughly dry, he melted them and submitted the material to cupellation, and we found three marks and a little more of fine silver. This he said had come from the cinnabar, the fine silver that we had put in having flown away in smoke. If this was profit, God knows it. And I also knew it, who had spent more than thirty crowns.

Several months later, Zachaire received a letter from the Abbé, asking him to return immediately, which he did. When he arrived at the abbey, he found that the King of Navarre had written to ask the Abbé to send him to the court at Pau to reveal the secret that the King understood him to possess. He was promised a reward of four thousand crowns.

The words, four thousand crowns, so tickled the ears of the Abbé that making himself believe that he had them already in his purse, he would not rest until I had started on my way to Pau. I arrived in the month of May and worked there for about six weeks ... and when I had finished I received the recompense I expected.

The four thousand crowns were in fact not forthcoming, but when the King dismissed him it was with a promise to reward him later.

This response annoyed me so much that without waiting for his lovely promises (having heretofore been nourished on them at my own expense) I started to return to the Abbé.

But having heard of a certain man learned in natural philosophy, he turned aside to ask his advice, which was that he should refrain from any further attempts at transmutation until he had gathered together and read as many of the books of the ancient philosophers as were obtainable. He then went on to meet the Abbé, to render an account of how the money had been spent – 'and to communicate to him half of the recompense that I had had from the King of Navarre'!

Afterwards he went on to his house, collected advance rent from his tenants, and returned to Paris, where he arrived the day following All Saints' in the year 1546.

There for ten crowns I bought books on philosophy, ancient as well as modern, part of which were printed while others were written by hand, the *Crowd of the Philosophers*, the good *Trevisan*, the *Complaint of Nature*, and other treatises which have never been printed. And having hired for myself a small room in the Faubourg Saint-Marceau, I lived there for a year, with a small boy who served me, without frequenting anyone, studying day and night on those authors, with the result that at the end of a month I made one conclusion, then another, then I changed it almost entirely. So, while waiting for a conclusion in which there was not variety nor contradiction between the sentences of the books of the philosophers, I spent an entire year and part of another without being able to gain over my study to the extent of being able to make any entire and perfect conclusion.

In his discouragement he began again to frequent the haunts of Parisian alchemists, but their operations seemed so chaotic and contradictory that he turned back to his books.

I began to re-read with very great diligence the works of Raymond Lully, principally his *Testament* and *Codicil*, which I reconciled so well with a letter which he had written in his time to King Robert [and with a recipe from another source] that I made a conclusion from them wholly contrary to all the operations which I had seen formerly but such that I read nothing in all the books which did not harmonize very well with my opinion, even the conclusions which Arnold of Villanova, who was master of Raymond Lully in this science, has made at the end of his *Great Rosary*. Thus I lived about one year longer without doing anything except read and think about my conclusion day and night, waiting until the term for which I had let my house should be passed to return to work at my own home.

Determined to practise the above-mentioned conclusion, I arrived there at the beginning of Lent, during which I provided myself with everything that I needed and set up a furnace for working, so that I commenced it the day after Easter. But this was not without having divers impediments [from relatives and friends, who urged the folly of throwing money away and of buying charcoal on such a scale as to arouse suspicion that counterfeiting was going on].

I leave it to you to imagine whether this talk was a bore to me, since at this time I was seeing my work go from better to better, and I was always attentive to the conduct of it in spite of these and comparable other delays which came upon me incessantly, and especially the danger of the plague, which was so great during the summer that there was no foot-travel or traffic which was not interrupted, in such manner that a day did not pass that I was not looking with very great diligence for the appearance of the three colours which the philosophers have written ought to appear before the perfection of our divine work. These, thanks to the Lord God, I saw, one after another, for on the very next Easter Day I saw the true and perfect experience of them on quicksilver heated in a crucible which I converted into fine gold under my own eyes in less than an hour by means of a little of this divine powder.

God knows if I was delighted about it. But I did not boast for all that. But after having rendered thanks to our good God who had shown me such grace and favour, I went away on the next day to find the Abbé at his monastery to fulfil the covenant and promise which we had made together. But I found that he had died six months previously, at which I was greatly grieved.

I therefore went away to a certain place to wait there for a friend of mine and near relative, who had lived with me at my residence and whom I had left there with authority and express instructions to sell all and each of the paternal goods which I had, to pay my creditors with the proceeds, and to distribute the rest secretly to those who were in need of it, in order that my relatives and others might feel some benefit from the great good that God had given me, without anyone being the wiser. But they thought on the contrary that, despairing and ashamed of my foolish expenditures, I had sold my goods in order to retire to another place, as this friend of mine informed me when he came to find me on the first day of July. And we went to Lausanne, I having decided to travel and to pass the rest of my days in a certain very renowned city of Germany, with a very small household, in order that I might not be known even by those who see and read this little book of mine during my lifetime, in our country of France.

Tenney L. Davis adds that 'the romantic ending which the eventful life of Denis Zachaire seems to require has been supplied, apparently out of the imagination, by Mardocheé de Delle, court poet of Rudolf II, who has told how our hero was married in Switzerland and started to travel. While stopping at Cologne, he was murdered in his sleep by a servant, who escaped with his wife and his store of transmuting powder. Figuier says that the event made considerable noise in Germany, but no traces of the assassin could be found.' Davis also says that the second part of the book, of which the autobiography forms the first, makes it clear that its author was well informed on the writings of the alchemists and on the methods of reasoning that they employed. It appears that he knew them so well that he was capable of satirizing them, for he presents faulty reasoning so plausibly that it appears preposterous. Indeed, he seems to have done for alchemy what Cervantes and Don Quixote did for knight-errantry.

Davis did not notice, or at least did not mention, the fact that a very similar story to Zachaire's is given in a work ascribed to Bernard of Treves (Trier), an alchemist about whom much confusion existed until Lynn Thorndike cleared up a good deal of it in 1934. We learn from Thorndike that in the second half of the fourteenth century an Italian physician, Thomas of Bologna, acted as astrologer and surgeon to Charles V of France, at a salary of one hundred francs a month. After Charles's death in 1380 he continued in the royal service under Charles VI, who in 1384 made him a grant of two hundred gold francs and spoke of him as 'Our beloved surgeon'. His daughter, Christine de Pisan, a poetess, historian, and feminist, says that her father was a doctor of medicine of Bologna University, where another authority relates that he taught astrology from 1345 to 1356.

One of his astrological exploits in France was directed against the bands of English mercenaries who were harassing the country. He obtained soil from the centre and four quarters of France and with it filled five leaden images of nude men, on which were written astrological characters and the name of the King of England or one of his captains. This procedure was carried out under selected constellations, and, again at the proper astrological moment, the images were buried face downwards in the places whence the soil had been taken. The burial was accompanied by incantation declaring that, as long as the images should last, the king and captains should suffer total destruction and annihilation, and that their mercenaries should likewise suffer perpetual expulsion and exclusion from France. Thomas and his royal patron must have been gratified at the result, for the narrator avers that within a few months all the said companies had fled from the realm.

But the successful astrologer had a bad moment when some medicines he prepared for the King and the Duke of Burgundy proved rather drastic in their effects and he was suspected of having given them something sinister. The affair blew over, but Thomas was worried about it, and wrote to Bernard of Treves asking him if he could discover whether his reputation had been seriously jeopardized. He explained to Bernard that the medicines had contained gold to which he had added mercury, which would have been in consonance with alchemical reasoning but might well have produced alarming symptoms in the patients. Perhaps to disguise his anxiety, Thomas did not limit his letter to these circumstances alone, but added a good deal of alchemical theorizing about his medicine, more efficacious and noble than all other medicines; in it golden solar sperm acted as a masculine agent on the feminine mercurial sperm. Bernard replied in a long letter, adopting a lofty but quite friendly tone, and the letter was afterwards printed several times, in Latin and in translations. He did not criticize Thomas's medicine as such, but said that it would be quite irrelevant to attempts at transmutation, and goes on to point out several misguided opinions in the alchemical theories that Thomas had added to his letter.

To come to Bernard himself: the dates formerly accepted for

his birth and death, namely 1406 and 1490, are clearly incorrect, since his contemporaneity with Thomas of Bologna places him definitely in the second half of the fourteenth century. Another error makes him Count of the March of Treviso, in Italy, while at the same time calling him a German doctor. In point of fact, at the end of his letter to Thomas, Bernard says that he completed it on St Denis's day, 1385, when he was for the time living at Treves; and on another occasion he addressed an alchemical tractate to the archbishop of that city. On the evidence, therefore, it appears that he was a fourteenth-century German alchemist.

The autobiography attributed to him resembles that of Zachaire so closely that we may wonder whether it was not a popular fiction fathered, with suitable variations, on more than one alchemist. Thus Bernard says that he spent four years and 400 crowns on testing a book by Razi, and then over 2000 crowns on books of Geber. Archelaus, Rupescissa, and John of Sacrobosco were other authorities whom he tested in vain, so that at the age of thirty-eight he had worn himself out with fruitless experimenting and was 6000 crowns the poorer. He then collaborated with an adept who had broken away from tradition and had the original idea of preparing the philosophers' stone from common salt dissolved by exposure to moist air and then crystallized in sunlight; this lost another eighteen months. The two of them became friendly with some other alchemists who were dissolving silver, copper, mercury, and other metals in acids. Each of the solutions had to stand for twelve months separately; then they were mixed and evaporated over hot ashes to two-thirds of the original bulk, the remaining solution being exposed to the sun's rays to crystallize. Twenty-two half-filled flasks were thus prepared, of which three were given to Bernard and his associate. Bernard waited for five years, that is until he was forty-six, but the liquids refused to crystallize, so he left his companion and joined forces with a monkish physician named Godfrey Leporis.

This partnership spent nearly ten years in manifold unsuccessful efforts, including one in which 2000 hen's eggs were used; and then, on the advice of a theologian, experimented with vitriol. The unfortunate result was that the fumes of the vitriol knocked Bernard out for fourteen months. His family now earnestly besought him to leave alchemy alone, and for a time he did so, but within a year he was again wandering far and wide in search of reliable information. He had been reduced to poverty and had had to sell his estate for 8000 German florins; altogether his researches had cost him over 10,000 crowns. At the aged of sixty-two, and in a very poor state of health, he arrived in Rhodes, where he prevailed upon a man of religion to lend him 8000 florins: these went in the same way as earlier ones. However, the man of religion had an extensive alchemical library, and for eight years, abstaining from experiment, Bernard read, studied, and pondered, and at last saw where he had gone wrong. He set to work again and was fully successful.

Christine de Pisan remembered her father's correspondence with Bernard, but she believed that all alchemists were deceived or deceivers. 'Such was one from Germany named Bernard, who kept great state and gained great renown and many followers; but in the end it was found that all was emptiness and deception.'

12

HELVETIUS, PRICE, AND SEMLER

ONE of the most circumstantial of all accounts of alleged transmutation comes from Johann Friedrich Schweitzer, generally known as Helvetius, who was born at Köthen, in the Duchy of Anhalt, in 1625, and died at The Hague in 1709. He was distinguished for his medical knowledge, and became physician to the Prince of Orange. While not going so far as to deny the pretensions of the alchemists, he was by no means convinced that they were justified, and he had no faith whatever in such absurdities as Digby's sympathetic powder (p. 213). A man of culture, education, and discernment, he can scarcely be suspected of having lied, or of wilfully misreporting the remarkable events he describes in the following narrative. In most accounts of 'transmutations' it is not difficult to perceive where trickery could have entered, but in the case of Helvetius no one has yet discovered the loophole. Even the nineteenth-century German chemist and historian of chemistry, Hermann Kopp, preferred to reserve judgement. Here is what Helvetius says:

The twenty seventh of December, 1666, in the afternoon, came a Stranger to my house at The Hague, in a plebeian habit, honest gravity, and serious authority; of a mean stature, a little long face, with a few small pock holes, and most black hair, not at all curled, a beardless chin, about three or four and forty years of age (as I guessed) and born in North Holland. After salutation he beseeched me with a great reverence to pardon his rude accesses, being a great lover of the Pyrotechnian Art; adding, he formerly endeavoured to visit me with a friend of his, and told me that he had read some of my small treatises; and particularly, that against the Sympathetick Powder of Sir Kenelm Digby, and observed my doubtfulness of the philosophical mystery, which caused him to take this opportunity, and asked me if I could not believe such a grand mystery was in nature, which could cure all diseases, unless the principal parts (as lungs, liver, &c.) were perished, or the predestinated time of death were come.

To which I replied, such a medicine would be a most desirable acquisition for any physician, but I never met with an Adept, or saw such a medicine, though I read much of it, and have wished for it. Then I asked if he were a physician, as he spoke so learnedly about the universal medicine, but he preventing my question, said he was a founder of brass, yet from his youth learnt many rare things in chymistry, of a friend, particularly the manner to extract out of metals many medicinal arcanas by force of fire, and was still a lover of it.

After other large discourse of experiments in metals this Elias asked me if I could know the philosophers' stone when I saw it, I answered not at all, though I had read much of it in Paracelsus, Helmont, Basilius, and others; yet I dare not say I could know the philosopher's matter. In the interim he took out of his bosom pouch or pocket, a cunningly-worked ivory box, and out of it took three ponderous pieces or small lumps of the stone, each about the bigness of a small walnut, transparent, of a pale brimstone colour, whereunto did stick the internal scales of the crucible, wherein it appeared this most noble substance was melted; they might be judged able to produce about twenty tons of gold, which when I had greedily seen and handled almost a quarter of a hour, and drawn from the owner many rare secrets of its admirable effects in human and metallic bodies, and other magical properties, I returned him this treasure of treasures; truly with a most sorrowful mind, after the custom of those who conquer themselves, yet (as was but just) very thankfully and humbly, I further desired to know why the colour was yellow, and not red, ruby colour, or purple, as philosophers write; he answered, that was nothing, for the matter was mature and ripe enough.

Then I humbly requested him to bestow a little piece of the medicine on me, in perpetual memory of him, though but the quantity of a coriander or hemp seed. He presently answered, O no, no, this is not lawful though thou wouldst give me as many ducats in gold as would fill this room, not for the value of the matter, but for some particular consequences not lawful to divulge, nay, if it were possible (said he) that fire could be burnt of fire, I would rather at this instant cast all this substance into the fiercest flames.

But after he demanding, if I had another private chamber, whose prospect was from the public street, I presently conducted him into the best furnished room backwards, where he enter'd without wiping his shoes (full of snow and dirt) according to the custom in Holland, then not doubting but he would bestow part thereof, or some great secret treasure on me, but in vain; for he asked for a little piece of gold and pulling off his cloak or pastoral habit, opened his doublet, under which he wore five pieces of gold hanging in green silk ribbons, as large as the inward round of a small pewter trencher: and this gold so far excelled mine, that there was no comparison, for flexibility and colour; and these figures [figure 10] with the inscriptions ingraven, were the resemblance, which he granted me to write out.

I being herewith affected with great admiration, desired to know where and how he came by them. He answered, An outlandish friend who dwelt some days in my house (giving out he was a lover of this art, and who came to reveal this art to me) taught me various arts: First, How out of ordinary stones and christalls, to make rubies, chrysolites, and sapphires, &c. much fairer than the ordinary. And how in a quarter of an hour to make crocus martis, of which one dose would infallibly cure the pestilential dysentery (or bloody flux), and how to make a metallic liquor most certainly to cure all kinds of dropsies in four days; as also a limpid clear water sweeter than honey, but which in two hours of itself, in hot sand, it would extract the tincture of granats, corals, glasses, and such like more, which I Helvetius did not observe, my mind being drawn beyond those bounds, to understand how such a noble juice might be drawn out of the metals, to transmute metals; but the shade in the water deceived the dog of the morsel of flesh in his mouth.

Moreover he told me his said master caused him to bring a glass full of warm rain water, and fetch some refined silver laminated into thin plates, which therein after he had added a little white powder was dissolved within a quarter of an hour, like ice when heated: And presently he drank to me the half, and I pledged him the other half, which had not so much taste as sweet milk; whereby methought I became very light hearted. I thereupon asked if this were a philosophical drink, and wherefore we drank this potion? He replied I ought not to be so curious. And after he told me that by the said master's directions, he took a piece of leaden pipe, gutter or cistern, and being melted put a little such sulphureous powder out of his Pocket and once again a little more on the point of a knife, and after a great blast of bellows in short time poured it on the red bricks of the kitchen chimney, which proved most excellent pure gold; which he said brought him into such a trembling amazement, that he could hardly speak. But his master thereupon again encouraged him, saying, cut for thyself the sixteenth part of this for a keepsake, and the rest give away amongst the poor, which he did. And he distributed so great an alms as he affirmed (if my memory

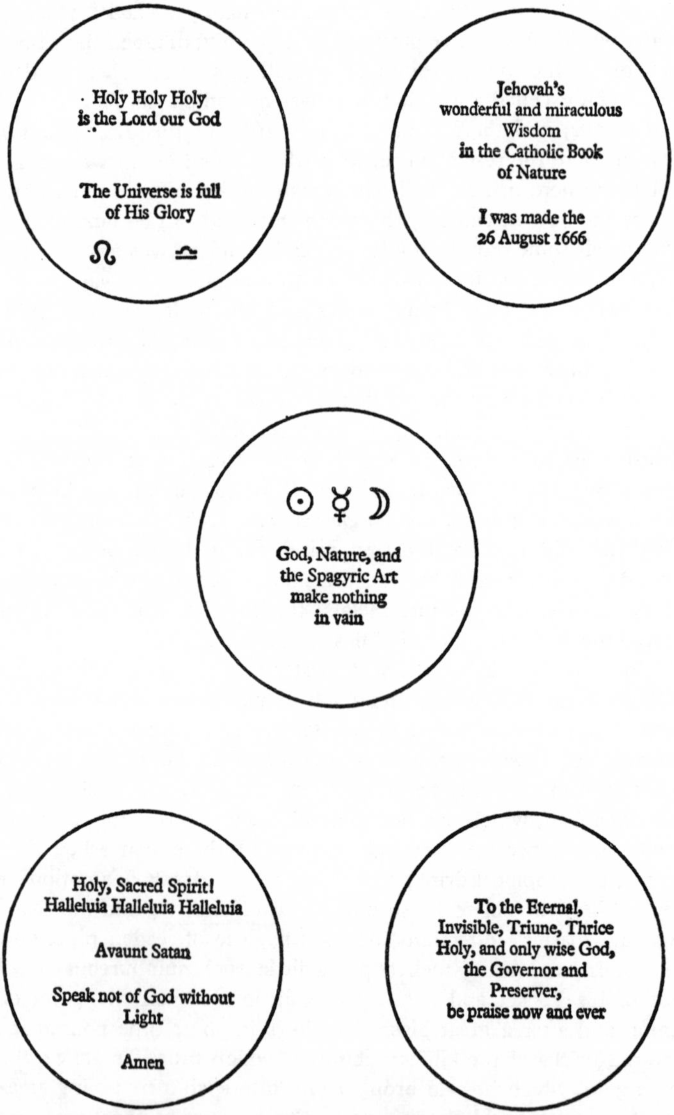

· Holy Holy Holy
is the Lord our God

The Universe is full
of His Glory

♌ ♎

Jehovah's
wonderful and miraculous
Wisdom
in the Catholic Book
of Nature

I was made the
26 August 1666

☉ ☿ ☽

God, Nature, and
the Spagyric Art
make nothing
in vain

Holy, Sacred Spirit!
Halleluia Halleluia Halleluia

Avaunt Satan

Speak not of God without
Light

Amen

To the Eternal,
Invisible, Triune, Thrice
Holy, and only wise God,
the Governor and
Preserver,
be praise now and ever

Figure 10. Alchemical Medals exhibited to Helvetius. (From J. F. Helvetius, *Vitulus Aureus*, Frankfort, 1767)

fail not) to the church of Sparrendam: but whether he gave it at several times or once, or in the golden masse, or in the silver coin, I did not ask.

At last said he (going on with the story of his master) he taught me thoroughly this Divine Art. As soon as his history was finished, I most humbly begged he would show me the effect of transmutation to confirm my faith therein, but he dismissed me for that time in such a discreet manner, that I had a denial. But withall promising to come again at three weeks end, and shew me some curious arts in the fire, and the manner of projection, provided it were then lawful without prohibition.

And at the three weeks end he came, and invited me abroad for an hour or two, and in our walks having discourses of divers of Nature's secrets in the fire; but he was very sparing in speaking of the great elixir, gravely asserting, that was only to magnify the sweet fame, and name of the most glorious God; and that few men endeavoured to sacrifice to him in good works, and this he expressed as a pastor or minister of a church; but now and then I kept his ears open, intreating to shew me the metallic transmutation; desiring also he would think me so worthy to eat and drink at my house, which I did prosecute so eagerly, that scarce any suitor could plead more to obtain his mistress from his corrival; but he was of so fixt and steadfast a spirit, that all my endeavours were frustrate: yet I could not forbear to tell him further I had a fit laboratory, and things ready and fit for an experiment, and that a promised favour was a kind of debt; yea, true said he, but I promised to teach thee at my return with this proviso, if it were not forbidden.

When I perceived all this in vain, I earnestly craved but a most small crumb or parcel of his powder or stone, to transmute four grains of lead to gold; and at last out of his philosophical commiseration, he gave me a crumb as big as a rape or turnip seed; saying, receive this small parcel of the greatest treasure in the world, which truly few kings or princes have ever known or seen: But I said, This perhaps will not transmute four grains of lead, whereupon he bid me deliver it him back, which in hopes of a greater parcel I did; but he cutting halfe off with his nail, flung it into the fire, and gave me the rest neatly wrapped in blue paper; saying, It is yet sufficient for thee. I answered him (indeed with a most dejected countenance) Sir, what means this? the other being too little, you give me now less.

He told me, If thou canst not manage this; yet for its great proportion for so small a quantity of lead, then put into the crucible two drams, or half an ounce, or a little more of the lead; for there ought to be no more lead put in the crucible than the medicine can work upon, and transmute. So I gave him great thanks for my diminished Treasure, concentrated truly in the superlative degree, and put the same charily up into my little box; saying I meant to try it the next day; nor would I reveal it to any.

Not so, not so; (said he) for we ought to divulge all things to the children of art; which may tend to the singular honour of God, that so they may live in theosophical truth, and not all die sophistically. After I made my confession to him, that whilst this mass of his medicine was in my hands, I endeavoured to scrape a little of it away with my nail, and could not forbear; but scratcht off nothing, or so very little, that it was but as an indivisible atom, which being purged from my nail, and wrapt in a paper: I projected on lead, but found no transmutation; but almost the whole masse of lead flew away, and the remainder turned into a meer glassy earth; at which unexpected passage, he smiling, said, thou art more dexterous to commit theft, than to apply thy Tincture; for if thou hadst only wrapped up thy stolen prey in yellow wax, to preserve it from the arising fumes of lead, it would have penetrated to the bottom of the lead, and transmuted it to gold; but having cast it into the fumes, partly by the violence of the vaporous fumes, and partly by the sympathetic alliance, it carried thy medicine quite away: For gold, silver, quicksilver, and the like metals, are corrupted and turn brittle like to glass, by the vapours of lead.

Whereupon I brought him my crucible wherein it was done, and instantly he perceived a most beautiful saffron-like tincture stick on the sides; and promised to come next morning, and then would shew me my error, and the said medicine should transmute the lead into gold. Nevertheless I earnestly prayed him in the interim to be pleased to declare only for my present instruction, if the philosophic work cost much, or required long time.

My friend, my friend (said he), thou art too curious to know all things in an instant, yet will I discover so much; that neither the great charge, or length of time, can discourage any; for as for the matter, out of which our magistery is made, I would have thee know there are only two metals and minerals, out of which it is prepared; but in regard the sulphur of philosophers is much more plentiful and abundant in the minerals; therefore it is made out of the minerals. Then I asked again, What was the menstruum, and whether the operation or working were done in glasses, or crucibles?

He answered, the menstruum was a heavenly salt, or of heavenly virtue, by whose benefit only the wise men dissolve the earthly metallic body, and by such a solution is easily and instantly brought forth the most noble elixir of philosophers. But in a crucible is all the operation done and performed, from the beginning to the very end, in an open fire, and all the whole work is no longer from the very first to last than four days, and the whole work is no more charge than three florins; and further, neither the mineral, out of which, nor the salt, by which it was performed, was of any great price. And when I replied, the philosophers affirm in their writings, that seven or nine months at least, are required for the work, he answered:

Their writings are only to be understood by the true adeptists; wherefore concerning time they would write nothing certain: Nay, without the communication of a true adept philosopher, not one student can find the way to prepare this great magistery, for which cause I warn and charge thee (as a friend) not to fling away thy money and goods to hunt out this art; for thou shalt never find it. To which I replied thy master (though unknown) shewed it thee; So mayst thou perchance discover something to me, that having overcome the rudiments, I may find the rest with little difficulty, according to the old saying, It is easier to add to a foundation, than begin a new.

He answered, In this art 'tis quite otherwise; for unless thou knowest the thing from the head to the heel, from the eggs to the apples; that is, from the very beginning to the very end, thou knowest nothing, and though I have told thee enough; yet thou knowest not how the philosophers do make, and break open the glassy seal of Hermes, in which the Sun sends forth a great splendour with his marvellous coloured metallic rayes, and in which looking-glass the eyes of Narcissus behold the transmutable metals, for out of those rays the true adept philosophers gather their fire; by whose help the volatile metals may be fixed into the most permanent metals, either gold or silver.

But enough at present; for I intend (God willing) once more tomorrow at the ninth hour (as I said) to meet, and discourse further on this philosophical subject, and shall shew you the manner of projection. And having taken his leave, he left me sorrowfully expecting him; but the next day he came not, nor ever since: only he sent an excuse at half an hour past nine that morning, by reason of his great business, and promised to come at three in the afternoon, but never came, nor have I heard of him since; whereupon I began to doubt of the whole matter.

Nevertheless late that night my wife (who was a most curious student and enquirer after the art, whereof that worthy man had discourst) came soliciting and vexing me to make experiment of that little spark of his bounty in that art, whereby to be the more assured of the truth; saying to me, unless this be done, I shall have no rest nor sleep all this night, but I wished her to have patience till next morning to expect this Elias; saying, perhaps he will return again to shew us the right manner.

In the meantime (she being so earnest) I commanded a fire to be made (thinking alas) now is this man (though so divine in discourse) found guilty of falsehood; and secondly attributing the error of my projecting the grand theft of his powder in the dirt of my nail to his charge, because it transmuted not the lead that time; and lastly, because he gave me too small a proportion of his said medicine (as I thought) to work upon so great a quantity of lead as he pretended and appointed for it. Saying further to myself, I fear, I fear indeed this man hath deluded me; nevertheless my wife wrapped the said matter in wax, and I cut half an ounce or six drams of old lead, and put it into a crucible in the fire, which being melted, my wife put in the said Medicine made up into a small pill or button, which presently made such a hissing and bubbling in its perfect operation, that within a quarter of an hour all the mass of lead was totally transmuted into the best and finest gold, which made us amazed as planet-struck.

And indeed (had I lived in Ovid's age) there could not have been a rarer metamorphosis than this, by the art of alchemy. Yea, could I have enjoyed Argus's eyes, with a hundred more, I could not sufficiently gaze upon this so admirable and almost miraculous a work of nature; for this melted lead (after projection) shewed us on the fire the rarest and most beautiful colours imaginable; yea, and the greenest colour, which as soon as poured forth into an ingot, it got the lively fresh colour of blood; and being cold shined as the purest and most refined resplendent gold. Truly I, and all standing about me, were exceedingly startled, and did run with this aurified lead (being yet hot) unto the goldsmith, who wondered at the fineness, and after a short trial of touch, he judg'd it the most excellent gold in the whole world, and offered to give most willingly fifty florins for every ounce of it.

The next day a rumour went about The Hague, and spread abroad; so that many illustrious persons and students gave me their friendly visits for its sake: Amongst the rest the general Assay-Master, or Examiner of the Coins of this Province of Holland, Mr Porelius, who with others earnestly beseeched me to pass some part of it through all their customary rituals, which I did, the rather to gratifie my own curiosity.

Thereupon we went to Mr Brechtel a silversmith, who first tried per quartam, viz. he mixt three or four parts of silver with one part of the said gold, and laminated, filed or granulated it, and put a sufficient quantity of Aqua Fort thereto, which presently dissolved the silver, and suffered the said gold to precipitate to the bottom, which being decanted off, and the calx or powder of gold dulcified with water, and then reduced and melted into a body, became excellent gold: And whereas we feared loss, we found that each dram of the said first gold was yet increased, and had transmuted a scruple of the said silver into gold, by reason of its great and excellent abounding tincture.

But now doubting further whether the silver was sufficiently separated from the said gold, we instantly mingled it with seven parts of antimony, which we melted and poured into a cone, and blowed off the regulus on a test, where we missed eight grains of our gold, but after we blowed away the rest of the antimony, or superfluous scoria, we found nine grains of gold more for our eight grains missing, yet this was somewhat pale and silver-like, which easily recovered its full colour afterwards. So that in the best proof of fire we lost nothing at all of this gold; but gained as aforesaid. The which proof I again repeated thrice, and the said remaining silver out of the aqua fortis, was of the very best flexible silver that could be; so that in the total, the said medicine (or elixir) had transmuted six drams and two scruples of the lead and silver, into most pure gold.

It is not surprising that a circumstantial narrative of this kind, related by a man of high standing, brought conviction to the minds of many doubters.

JAMES PRICE

Another interesting story, though with a less satisfactory conclusion, comes from England in the last quarter of the eighteenth century. The central figure in it is James Higginbotham, who was born in London in 1752. He matriculated at Magdalen Hall, Oxford, in 1772, and took his M.A. degree in 1777 and M.D. in 1782. In 1781 he changed his name to Price, in order to comply with the wishes of a relative who had bequeathed him a large sum of money, and in the same year he was elected a Fellow of the Royal Society. His certificate of recommendation to the Society described him as 'well versed in various branches of Natural Science, and particularly in Chymistry'; it seems also that his doctor's degree was awarded for his proficiency in chemistry rather than medicine. The 'learned, ingenious, and pleasing' Dr Wall, who once drank tea with Samuel Johnson, thought highly of Price's chemical ability, and it was largely due to Wall's influence that Price so quickly rose to a position of some eminence in the scientific world.

On the strength of his legacy, Price went down from Oxford in 1782 and settled in a country house at Stoke, near Guildford in Surrey, and there built and equipped a chemical laboratory. He had not long been at Stoke when he announced that he had succeeded in transmuting mercury into gold and invited some distinguished men, including Lords Onslow, Palmerston, and King, to his laboratory to witness the experiment. None of the witnesses had had any training in science and they were very much impressed when Price added a white powder to fifty times its own weight of mercury, mixed them with a flux of borax and nitre, and heated them in a crucible. When the cooled mass was turned out of the crucible it was seen to be an ingot of silver equal in weight to the mercury used. A similar operation was carried out with a red powder and sixty times its weight of mercury; in this case the product was gold. Price would not divulge the composition of his powders, nor the way in which he prepared them, but he agreed that the specimens of silver and gold so obtained should be submitted to assay. They proved to be genuine, and were exhibited to George III.

Needless to say these experiments, which Price described in a pamphlet published shortly afterwards, created a great sensation; they were quoted at some length in the *London Chronicle* and on the Continent, and the pamphlet itself was translated into German. Price brought out a second edition of the pamphlet in 1783, saying that while his experiments were incontestable he nevertheless did not believe in the philosophers' stone, which he regarded as a chimera; he added in the preface that his stock of the powders was now exhausted, and that the cost of making a further supply, especially in its demands on health and labour, would be too great for him to undertake it.

Meanwhile a correspondent had written to the famous chemist Joseph Black (1728–99) expressing the view that Price's claims were nothing but a mass of errors throughout. Black knew that Price's advancement in science had been speeded by Wall's recommendation and in his reply said that he had in fact been very much surprised that Price should have ever been given the degree of M.D.; as for the pamphlet, he thought that it was necessary to take some notice of it in his introductory lecture to his university students that season, 'but my only intention in this was earnestly to dissuade my pupils from giving way in the least to the ruinous notions and pursuits of Alchemy, which Dr P's publication, so far as it may be credited, would have a tendency to encourage. I reminded my audience that numerous experiments to prove the possibility of transformation had been formerly exhibited or described which were afterwards discovered to have been founded on mistakes or frauds: and observed that the present age does not require incentive to experimental inquiries, with a view to deterring my young friends from entering Alchemical Notions or Projects.'

D. I. Duveen, who quotes this letter of Black's, adds that Wall later excused himself for having supported Price and said 'Whatever other defects we may have, an attachment to the hermetic Philosophy makes no part of our character.'

However, Price's pamphlets had aroused so much controversy that the Royal Society felt it imperative to have the matter settled. The President, Sir Joseph Banks, reminded him that the honour of the Society as well as his own was at stake, and that the experiments should be repeated before officially appointed Fellows as witnesses. For some time Price refused to obey, but his friends put such pressure on him that in the end he agreed to prepare further quantities of the powders and to have them ready in six weeks. Rather desperately, he tried to get alchemical information from Germany; but the six weeks had nearly expired. Early in August the three delegates of the Society arrived at Stoke and Price conducted them to his laboratory. Leaving them for a moment, he drank a concoction containing hydrocyanic acid (prussic acid), returned to the laboratory, and collapsed and died before their eyes. The question whether Price was led to imposture by vanity or madness, or whether the mercury of his original experiments had been deliberately adulterated by dealers who knew of his alchemical beliefs and wanted to encourage them to their own profit, cannot now be solved. At the inquest, Price was found to have been of unsound mind, a verdict that may well represent the truth, since he was certainly a competent chemist and would therefore probably not have been duped by impure mercury.

JOHANN SEMLER

The case of Price was a tragic one, but Figuier relates another of the same period which provides light relief. It concerns a learned German scholar, Johann Semler, who was professor of theology in the university of Halle. As a child, Semler had often heard an alchemist friend of his father describe the wonderful properties of the Stone, and though he became a theologian he practised chemistry in his hours of leisure. It is true that he could not prepare anything even remotely resembling the marvellous substance that had so fired his childish imagination, but he never disbelieved in the possibility of its existence. His faith was confirmed by his profound study of the alchemical classics.

It so happened that in 1786 Baron Leopold von Hirschen announced that he had discovered a substance with remarkable curative powers, which he called the 'salt of life'. Semler enthusiastically undertook the investigation of this salt and wrote three memoirs on it. He not merely confirmed von Hirschen's claims but improved upon them: the salt was more than an ordinary medicine, it was the panacea for all diseases and, still more astonishing, it could cause metallic transmutation. For the latter purpose, none of the usual paraphernalia were necessary – no fire, no crucible, no sulphur or mercury: all that had to be done was to dissolve the salt in water and leave the solution for some days in a flask maintained at somewhat higher than room temperature. Under these conditions, gold would gradually be deposited at the bottom of the vessel.

Semler's position as holder of an important university chair prevented his assertions from being rejected out of hand; no one could doubt his sincerity, and that he was convinced of the truth of his claims was not questioned. He was, however, asked to provide experimental proof, a request to which he readily agreed. The chemist Friedrich Gren undertook to examine the salt, of which Semler sent him a sample; but as soon as Gren looked at it carefully he saw that it was mixed with bits of gold leaf. Semler agreed that the gold was there, but maintained that it had arisen spontaneously. To decide the matter, it was agreed that the salt should be submitted to Heinrich Klaproth, professor of chemistry at Berlin and one of the leading chemists of his day. Klaproth conducted an analysis of the salt, and found that it consisted of a mixture of Glauber's salt, Epsom salt, and bits of gold leaf, made into a paste with urine; but wishing to place the matter beyond doubt he asked Semler to send him a further sample. The result of analysis was exactly the same as before, and Klaproth was convinced of the stupidity of the whole affair. Semler, however, persisted. He wrote again to Klaproth, saying 'My experiments are well advanced. Two of my flasks are bearing gold; I take it out every five or six days, and each time I withdraw about twelve to fifteen grains. Two or three other flasks are coming along well; leaves of gold can already be seen at the bottom. But all this, so far, is costing me a great deal, for to get a grain of gold needs an outlay of two to four thalers; perhaps that is because I am not yet sufficiently skilful in the manner of operating.' So sure was he of his ground that he begged Klaproth to conduct still another analysis, on this occasion before a public audience.

Klaproth doubtless felt that that might be the only way of refuting Semler's claim, and the meeting was arranged with a good deal of publicity. Eminent persons, including ministers of the Crown, thronged the hall, and Berlin waited with impatience for the result of the tests. It was a staggering one, for the very first reagent that Klaproth applied to the 'gold' proved that it was not gold at all, but brass. A huge roar of laughter went up from the assembled company, and was echoed throughout Germany. At last the professor of Halle realized that someone had been deceiving him, and he started to investigate. It then came to light that his manservant, who was very much attached to his master, and desired nothing more than that the professor's experiments should succeed, had been in the habit of surreptitiously adding fragments of gold leaf to the flasks containing the 'salt of life', but he had recently been called up for a period of military training and had entrusted his wife with the task. This she dutifully performed, but her thrifty mind was shocked at what she considered to be a waste of expensive gold leaf. Brass foil had the same appearance and cost much less, so why not use that instead?

Thus the whole story came out. Semler made a frank admission that he had been duped, and after the general amusement had died away he regained public esteem, but in the realm of theology rather than in that of chemistry.

EPILOGUE

AFTER the Royal Society's investigation of James Price's assertions, no learned scientific body has been willing officially to notice alchemical claims. By the end of the seventeenth century a new chemistry had begun to take shape, in which the basic conceptions and theories of alchemy had no part. The founder of this rational system, from which our science of chemistry has arisen, was the Hon. Robert Boyle (1627–91), who in 1661 published his – literally – epoch-making book entitled *The Sceptical Chymist*. In this book he struck at the root of all alchemical speculation by denying that the four Aristotelian 'elements' had any right whatever to that description. Elements, he said, should be regarded as 'primitive and simple, or perfectly unmingled bodies; which not being made of any other bodies, or of one another, are the ingredients of which all those called perfectly mixt bodies are immediately compounded, and into which they are ultimately resolved ... I must not look upon any body as a true principle or element, which is not perfectly homogeneous, but is further resolvable into any number of distinct substances'. In other words, chemists should regard as elementary all those substances that they have not yet been able to split up into two or more constituents, and should not limit themselves by any preconceived notions of the number of these elements. If a substance is undecomposable it is to be considered an element, and it will retain that status for just so long as it withstands the efforts of chemists to decompose it.

The publication of *The Sceptical Chymist* was the death-warrant of alchemy, though, as we have seen, it survived in apparent vigour for another century. Boyle's demolition of the four elements was followed about eighty years later by Joseph Black's introduction of quantitative methods in chemical research, and not long afterwards by Priestley's discovery of oxygen and Lavoisier's elucidation of the composition of air and water. A chemical revolution had taken place, and was finally established by Dalton's atomic theory published in the opening decade of the nineteenth century. Chemists now had in view wider and more promising fields than those cultivated without success by the alchemists. Yet it must be remembered that to the alchemists was due much of the practical chemical knowledge upon which scientific chemistry was based; and until the reorientation initiated by Boyle there was nothing strange or inconsistent in the idea that one metal might be converted into another.

Perhaps we may leave the last words to Boerhaave (1668–1738), a Dutch chemist and physician 'who, by his renown in Europe and by the extent of his knowledge, by himself alone was worth almost a whole academy'. In his *New Method of Chemistry* he wrote that if he were asked his opinion about the philosophers' stone,

I should answer, that the wise Socrates, after reading a most abstruse book of Heraclitus, being ask'd what he thought of it, replied, that where he understood it, he found it excellent, and believ'd it to be so in those other parts he could not comprehend, which required the greatest penetration to come at. So wherever I understand the alchemists, I find them describe the truth in the most simple and naked terms, without deceiving us, or being deceived themselves. When therefore I come to places, where I do not comprehend the meaning; why should I charge them with falsehood, who have shown themselves so much better skill'd in the art than myself; from whom I have learnt many things, in those parts of their writings where they thought proper to speak plain? ... I therefore rather lay the blame on my own ignorance than on their vanity. Yet I have often doubted, upon reading their secrets, whether these skilful persons, after they had discovered so many extraordinary things by naked observations, might not by a too great quickness of apprehension anticipate, and relate things for facts, which they conclude might be done; or even must of necessity have been done, if they had persisted in the pursuit. 'Tis certain a very grave alchemist, Alexander Suchthen [Seton], a disciple of Paracelsus, and a zealot in defending his doctrine, had tried so many things to so little purpose, that he concludes, at the end of his treatise of antimony, that all the philosophers, the principal of whom he there recites, had died before they brought their speculations to an issue. If this be the case, which I shall not pretend to determine, we are nevertheless exceedingly obliged to them for the immense pains they have been at, in discovering, and handing to us, so many difficult physical truths: Insomuch, that Lord Bacon justly compares them to a father, who on his death-bed inform'd his lazy sons, of a sum of money which he had hid underground in his garden. After his death they went to digging, in hopes of finding the treasure, and tho' they missed their aim, for in reality there was none hid, yet they sufficiently enrich'd themselves, by the large crop which the ground, in consequence of this tillage, produced. Thus much I have long ago had a mind to say, concerning the knowledge of the true alchemists in physics; lest such skilful artists should be condemn'd by incompetent judges.... Credulity is hurtful, so is incredulity: the business therefore of a wise man is to try all things, hold fast what is approv'd, never limit the power of God, nor assign bounds to nature.

GLOSSARY

Alembic, still-head or complete still.

Aludel, condensing receiver used to collect sublimates.

Aqua fortis, nitric acid.

Aqua regia, a mixture of nitric acid and hydrochloric acid; it will dissolve gold.

Argentvive, mercury, quicksilver.

Athanor, furnace (Arabic *al-tannur*).

Azoth, mercury, especially the hypothetical mercury supposed to be the first principle of metals.

Bain-marie, water-bath.

Caduceus, wand of Hermes or other messenger.

Calcination, conversion of a metal or other mineral to the state of a fine powder, typically by means of heat.

Calx, product of calcination.

Ceration, conversion of a substance to a waxy condition.

Cinnabar, ore of mercury, mercuric sulphide, HgS.

Circulation, reflux distillation.

Coagulation, crystallization, conversion of a liquid to a solid.

Cohobation, return of a liquid distillate to its residue followed by further distillation, and so on.

Congelation, crystallization.

Croslet, crucible.

Cupel, crucible made of bone-ash for testing or refining gold and silver.

Curcurbit(e), flask or "gourd" forming lower part of a still.

Descension, distillation or fusion in which liquid product flows down into a receiver below.

Distillation, boiling a liquid and reconverting the vapour into a liquid by cooling; sometimes applied to the removal of liquid from one vessel to another by capillary movement through a wick or strip of cloth.

Ferment, yeast barm; sometimes applied to the philosophers' stone.

Fixation, rendering a substance non-volatile.

Hypostatical principles, sulphur, mercury, and salt.

Imbibition, same as cohobation.

Multiplication, transmutation, alchemical preparation of gold, artificially induced growth of gold.

Pelican, form of circulatory still.

Projection, addition of the philosophers' stone to the material prepared for transmutation.

Puffers, alchemists, from their excessive use of bellows.

Putrefaction, conversion of a metal into an apparently inert mass or powder.

Red work, conversion of base metals into gold.

Sophic hydrolith, philosophers' stone.

Souffleurs, see *Puffers*.

Tincture, philosophers' stone or its action upon base metals.

White water, mercury.

White work, conversion of base metals into silver.

INDEX OF PERSONS

GENERAL INDEX

A CATALOG OF SELECTED

DOVER BOOKS

IN ALL FIELDS OF INTEREST

A CATALOG OF SELECTED DOVER BOOKS IN ALL FIELDS OF INTEREST

THE BOOK OF BEASTS: Being a Translation from a Latin Bestiary of the Twelfth Century, T. H. White. Wonderful catalog real and fanciful beasts: manticore, griffin, phoenix, amphivius, jaculus, many more. White's witty erudite commentary on scientific, historical aspects. Fascinating glimpse of medieval mind. Illustrated. 296pp. 5⅜ × 8¼. (Available in U.S. only) 24609-4 Pa. $6.95

FRANK LLOYD WRIGHT: ARCHITECTURE AND NATURE With 160 Illustrations, Donald Hoffmann. Profusely illustrated study of influence of nature—especially prairie—on Wright's designs for Fallingwater, Robie House, Guggenheim Museum, other masterpieces. 96pp. 9¼ × 10¾. 25098-9 Pa. $8.95

FRANK LLOYD WRIGHT'S FALLINGWATER, Donald Hoffmann. Wright's famous waterfall house: planning and construction of organic idea. History of site, owners, Wright's personal involvement. Photographs of various stages of building. Preface by Edgar Kaufmann, Jr. 100 illustrations. 112pp. 9¼ × 10.
23671-4 Pa. $8.95

YEARS WITH FRANK LLOYD WRIGHT: Apprentice to Genius, Edgar Tafel. Insightful memoir by a former apprentice presents a revealing portrait of Wright the man, the inspired teacher, the greatest American architect. 372 black-and-white illustrations. Preface. Index. vi + 228pp. 8¼ × 11. 24801-1 Pa. $10.95

THE STORY OF KING ARTHUR AND HIS KNIGHTS, Howard Pyle. Enchanting version of King Arthur fable has delighted generations with imaginative narratives of exciting adventures and unforgettable illustrations by the author. 41 illustrations. xviii + 313pp. 6⅛ × 9¼. 21445-1 Pa. $6.95

THE GODS OF THE EGYPTIANS, E. A. Wallis Budge. Thorough coverage of numerous gods of ancient Egypt by foremost Egyptologist. Information on evolution of cults, rites and gods; the cult of Osiris; the Book of the Dead and its rites; the sacred animals and birds; Heaven and Hell; and more. 956pp. 6⅛ × 9¼.
22055-9, 22056-7 Pa., Two-vol. set $21.90

A THEOLOGICO-POLITICAL TREATISE, Benedict Spinoza. Also contains unfinished *Political Treatise*. Great classic on religious liberty, theory of government on common consent. R. Elwes translation. Total of 421pp. 5⅜ × 8½.
20249-6 Pa. $7.95

INCIDENTS OF TRAVEL IN CENTRAL AMERICA, CHIAPAS, AND YUCATAN, John L. Stephens. Almost single-handed discovery of Maya culture; exploration of ruined cities, monuments, temples; customs of Indians. 115 drawings. 892pp. 5⅜ × 8½. 22404-X, 22405-8 Pa., Two-vol. set $15.90

LOS CAPRICHOS, Francisco Goya. 80 plates of wild, grotesque monsters and caricatures. Prado manuscript included. 183pp. 6⅝ × 9⅝. 22384-1 Pa. $5.95

AUTOBIOGRAPHY: The Story of My Experiments with Truth, Mohandas K. Gandhi. Not hagiography, but Gandhi in his own words. Boyhood, legal studies, purification, the growth of the Satyagraha (nonviolent protest) movement. Critical, inspiring work of the man who freed India. 480pp. 5⅜ × 8½. (Available in U.S. only)
24593-4 Pa. $6.95

ILLUSTRATED DICTIONARY OF HISTORIC ARCHITECTURE, edited by Cyril M. Harris. Extraordinary compendium of clear, concise definitions for over 5,000 important architectural terms complemented by over 2,000 line drawings. Covers full spectrum of architecture from ancient ruins to 20th-century Modernism. Preface. 592pp. 7½ × 9⅜. 24444-X Pa. $15.95

THE NIGHT BEFORE CHRISTMAS, Clement Moore. Full text, and woodcuts from original 1848 book. Also critical, historical material. 19 illustrations. 40pp. 4⅝ × 6. 22797-9 Pa. $2.50

THE LESSON OF JAPANESE ARCHITECTURE: 165 Photographs, Jiro Harada. Memorable gallery of 165 photographs taken in the 1930's of exquisite Japanese homes of the well-to-do and historic buildings. 13 line diagrams. 192pp. 8⅜ × 11¼. 24778-3 Pa. $10.95

THE AUTOBIOGRAPHY OF CHARLES DARWIN AND SELECTED LETTERS, edited by Francis Darwin. The fascinating life of eccentric genius composed of an intimate memoir by Darwin (intended for his children); commentary by his son, Francis; hundreds of fragments from notebooks, journals, papers; and letters to and from Lyell, Hooker, Huxley, Wallace and Henslow. xi + 365pp. 5⅜ × 8. 20479-0 Pa. $6.95

WONDERS OF THE SKY: Observing Rainbows, Comets, Eclipses, the Stars and Other Phenomena, Fred Schaaf. Charming, easy-to-read poetic guide to all manner of celestial events visible to the naked eye. Mock suns, glories, Belt of Venus, more. Illustrated. 299pp. 5¼ × 8¼. 24402-4 Pa. $7.95

BURNHAM'S CELESTIAL HANDBOOK, Robert Burnham, Jr. Thorough guide to the stars beyond our solar system. Exhaustive treatment. Alphabetical by constellation: Andromeda to Cetus in Vol. 1; Chamaeleon to Orion in Vol. 2; and Pavo to Vulpecula in Vol. 3. Hundreds of illustrations. Index in Vol. 3. 2,000pp. 6⅛ × 9¼. 23567-X, 23568-8, 23673-0 Pa., Three-vol. set $41.85

STAR NAMES: Their Lore and Meaning, Richard Hinckley Allen. Fascinating history of names various cultures have given to constellations and literary and folkloristic uses that have been made of stars. Indexes to subjects. Arabic and Greek names. Biblical references. Bibliography. 563pp. 5⅜ × 8½. 21079-0 Pa. $8.95

THIRTY YEARS THAT SHOOK PHYSICS: The Story of Quantum Theory, George Gamow. Lucid, accessible introduction to influential theory of energy and matter. Careful explanations of Dirac's anti-particles, Bohr's model of the atom, much more. 12 plates. Numerous drawings. 240pp. 5⅜ × 8½. 24895-X Pa. $5.95

CHINESE DOMESTIC FURNITURE IN PHOTOGRAPHS AND MEASURED DRAWINGS, Gustav Ecke. A rare volume, now affordably priced for antique collectors, furniture buffs and art historians. Detailed review of styles ranging from early Shang to late Ming. Unabridged republication. 161 black-and-white drawings, photos. Total of 224pp. 8⅜ × 11¼. (Available in U.S. only) 25171-3 Pa. $13.95

VINCENT VAN GOGH: A Biography, Julius Meier-Graefe. Dynamic, penetrating study of artist's life, relationship with brother, Theo, painting techniques, travels, more. Readable, engrossing. 160pp. 5⅜ × 8½. (Available in U.S. only) 25253-1 Pa. $4.95

HOW TO WRITE, Gertrude Stein. Gertrude Stein claimed anyone could understand her unconventional writing—here are clues to help. Fascinating improvisations, language experiments, explanations illuminate Stein's craft and the art of writing. Total of 414pp. 4⅜ × 6⅜. 23144-5 Pa. $6.95

ADVENTURES AT SEA IN THE GREAT AGE OF SAIL: Five Firsthand Narratives, edited by Elliot Snow. Rare true accounts of exploration, whaling, shipwreck, fierce natives, trade, shipboard life, more. 33 illustrations. Introduction. 353pp. 5⅜ × 8½. 25177-2 Pa. $8.95

THE HERBAL OR GENERAL HISTORY OF PLANTS, John Gerard. Classic descriptions of about 2,850 plants—with over 2,700 illustrations—includes Latin and English names, physical descriptions, varieties, time and place of growth, more. 2,706 illustrations. xlv + 1,678pp. 8½ × 12¼. 23147-X Cloth. $75.00

DOROTHY AND THE WIZARD IN OZ, L. Frank Baum. Dorothy and the Wizard visit the center of the Earth, where people are vegetables, glass houses grow and Oz characters reappear. Classic sequel to *Wizard of Oz*. 256pp. 5⅜ × 8. 24714-7 Pa. $5.95

SONGS OF EXPERIENCE: Facsimile Reproduction with 26 Plates in Full Color, William Blake. This facsimile of Blake's original "Illuminated Book" reproduces 26 full-color plates from a rare 1826 edition. Includes "The Tyger," "London," "Holy Thursday," and other immortal poems. 26 color plates. Printed text of poems. 48pp. 5¼ × 7. 24636-1 Pa. $3.95

SONGS OF INNOCENCE, William Blake. The first and most popular of Blake's famous "Illuminated Books," in a facsimile edition reproducing all 31 brightly colored plates. Additional printed text of each poem. 64pp. 5¼ × 7. 22764-2 Pa. $3.95

PRECIOUS STONES, Max Bauer. Classic, thorough study of diamonds, rubies, emeralds, garnets, etc.: physical character, occurrence, properties, use, similar topics. 20 plates, 8 in color. 94 figures. 659pp. 6⅛ × 9¼. 21910-0, 21911-9 Pa., Two-vol. set $15.90

ENCYCLOPEDIA OF VICTORIAN NEEDLEWORK, S. F. A. Caulfeild and Blanche Saward. Full, precise descriptions of stitches, techniques for dozens of needlecrafts—most exhaustive reference of its kind. Over 800 figures. Total of 679pp. 8⅛ × 11. Two volumes. Vol. 1 22800-2 Pa. $11.95
Vol. 2 22801-0 Pa. $11.95

THE MARVELOUS LAND OF OZ, L. Frank Baum. Second Oz book, the Scarecrow and Tin Woodman are back with hero named Tip, Oz magic. 136 illustrations. 287pp. 5⅜ × 8½. 20692-0 Pa. $5.95

WILD FOWL DECOYS, Joel Barber. Basic book on the subject, by foremost authority and collector. Reveals history of decoy making and rigging, place in American culture, different kinds of decoys, how to make them, and how to use them. 140 plates. 156pp. 7⅞ × 10¾. 20011-6 Pa. $8.95

HISTORY OF LACE, Mrs. Bury Palliser. Definitive, profusely illustrated chronicle of lace from earliest times to late 19th century. Laces of Italy, Greece, England, France, Belgium, etc. Landmark of needlework scholarship. 266 illustrations. 672pp. 6⅛ × 9¼. 24742-2 Pa. $14.95

SUNDIALS, Albert Waugh. Far and away the best, most thorough coverage of ideas, mathematics concerned, types, construction, adjusting anywhere. Over 100 illustrations. 230pp. 5⅜ × 8½. 22947-5 Pa. $5.95

PICTURE HISTORY OF THE NORMANDIE: With 190 Illustrations, Frank O. Braynard. Full story of legendary French ocean liner: Art Deco interiors, design innovations, furnishings, celebrities, maiden voyage, tragic fire, much more. Extensive text. 144pp. 8⅞ × 11¾. 25257-4 Pa. $10.95

THE FIRST AMERICAN COOKBOOK: A Facsimile of "American Cookery," 1796, Amelia Simmons. Facsimile of the first American-written cookbook published in the United States contains authentic recipes for colonial favorites—pumpkin pudding, winter squash pudding, spruce beer, Indian slapjacks, and more. Introductory Essay and Glossary of colonial cooking terms. 80pp. 5⅜ × 8½. 24710-4 Pa. $3.50

101 PUZZLES IN THOUGHT AND LOGIC, C. R. Wylie, Jr. Solve murders and robberies, find out which fishermen are liars, how a blind man could possibly identify a color—purely by your own reasoning! 107pp. 5⅜ × 8½. 20367-0 Pa. $2.50

ANCIENT EGYPTIAN MYTHS AND LEGENDS, Lewis Spence. Examines animism, totemism, fetishism, creation myths, deities, alchemy, art and magic, other topics. Over 50 illustrations. 432pp. 5⅜ × 8½. 26525-0 Pa. $8.95

ANTHROPOLOGY AND MODERN LIFE, Franz Boas. Great anthropologist's classic treatise on race and culture. Introduction by Ruth Bunzel. Only inexpensive paperback edition. 255pp. 5⅜ × 8½. 25245-0 Pa. $6.95

THE TALE OF PETER RABBIT, Beatrix Potter. The inimitable Peter's terrifying adventure in Mr. McGregor's garden, with all 27 wonderful, full-color Potter illustrations. 55pp. 4¼ × 5½. (Available in U.S. only) 22827-4 Pa. $1.75

THREE PROPHETIC SCIENCE FICTION NOVELS, H. G. Wells. *When the Sleeper Wakes, A Story of the Days to Come* and *The Time Machine* (full version). 335pp. 5⅜ × 8½. (Available in U.S. only) 20605-X Pa. $6.95

APICIUS COOKERY AND DINING IN IMPERIAL ROME, edited and translated by Joseph Dommers Vehling. Oldest known cookbook in existence offers readers a clear picture of what foods Romans ate, how they prepared them, etc. 49 illustrations. 301pp. 6⅛ × 9¼. 23563-7 Pa. $7.95

SHAKESPEARE LEXICON AND QUOTATION DICTIONARY, Alexander Schmidt. Full definitions, locations, shades of meaning of every word in plays and poems. More than 50,000 exact quotations. 1,485pp. 6½ × 9¼. 22726-X, 22727-8 Pa., Two-vol. set $31.90

THE WORLD'S GREAT SPEECHES, edited by Lewis Copeland and Lawrence W. Lamm. Vast collection of 278 speeches from Greeks to 1970. Powerful and effective models; unique look at history. 842pp. 5⅜ × 8½. 20468-5 Pa. $12.95

PLANTS OF THE BIBLE, Harold N. Moldenke and Alma L. Moldenke. Standard reference to all 230 plants mentioned in Scriptures. Latin name, biblical reference, uses, modern identity, much more. Unsurpassed encyclopedic resource for scholars, botanists, nature lovers, students of Bible. Bibliography. Indexes. 123 black-and-white illustrations. 384pp. 6 × 9. 25069-5 Pa. $8.95

FAMOUS AMERICAN WOMEN: A Biographical Dictionary from Colonial Times to the Present, Robert McHenry, ed. From Pocahontas to Rosa Parks, 1,035 distinguished American women documented in separate biographical entries. Accurate, up-to-date data, numerous categories, spans 400 years. Indices. 493pp. 6½ × 9¼. 24523-3 Pa. $10.95

THE FABULOUS INTERIORS OF THE GREAT OCEAN LINERS IN HISTORIC PHOTOGRAPHS, William H. Miller, Jr. Some 200 superb photographs capture exquisite interiors of world's great "floating palaces"—1890's to 1980's: *Titanic, Ile de France, Queen Elizabeth, United States, Europa,* more. Approx. 200 black-and-white photographs. Captions. Text. Introduction. 160pp. 8⅝ × 11¾. 24756-2 Pa. $9.95

THE GREAT LUXURY LINERS, 1927–1954: A Photographic Record, William H. Miller, Jr. Nostalgic tribute to heyday of ocean liners. 186 photos of Ile de France, Normandie, Leviathan, Queen Elizabeth, United States, many others. Interior and exterior views. Introduction. Captions. 160pp. 9 × 12. 24056-8 Pa. $10.95

A NATURAL HISTORY OF THE DUCKS, John Charles Phillips. Great landmark of ornithology offers complete detailed coverage of nearly 200 species and subspecies of ducks: gadwall, sheldrake, merganser, pintail, many more. 74 full-color plates, 102 black-and-white. Bibliography. Total of 1,920pp. 8⅝ × 11¼. 25141-1, 25142-X Cloth. Two-vol. set $100.00

THE SEAWEED HANDBOOK: An Illustrated Guide to Seaweeds from North Carolina to Canada, Thomas F. Lee. Concise reference covers 78 species. Scientific and common names, habitat, distribution, more. Finding keys for easy identification. 224pp. 5⅜ × 8½. 25215-9 Pa. $6.95

THE TEN BOOKS OF ARCHITECTURE: The 1755 Leoni Edition, Leon Battista Alberti. Rare classic helped introduce the glories of ancient architecture to the Renaissance. 68 black-and-white plates. 336pp. 8⅜ × 11¼. 25239-6 Pa. $14.95

MISS MACKENZIE, Anthony Trollope. Minor masterpieces by Victorian master unmasks many truths about life in 19th-century England. First inexpensive edition in years. 392pp. 5⅜ × 8½. 25201-9 Pa. $8.95

THE RIME OF THE ANCIENT MARINER, Gustave Doré, Samuel Taylor Coleridge. Dramatic engravings considered by many to be his greatest work. The terrifying space of the open sea, the storms and whirlpools of an unknown ocean, the ice of Antarctica, more—all rendered in a powerful, chilling manner. Full text. 38 plates. 77pp. 9¼ × 12. 22305-1 Pa. $4.95

THE EXPEDITIONS OF ZEBULON MONTGOMERY PIKE, Zebulon Montgomery Pike. Fascinating first-hand accounts (1805–6) of exploration of Mississippi River, Indian wars, capture by Spanish dragoons, much more. 1,088pp. 5⅜ × 8½. 25254-X, 25255-8 Pa. Two-vol. set $25.90

A CONCISE HISTORY OF PHOTOGRAPHY: Third Revised Edition, Helmut Gernsheim. Best one-volume history—camera obscura, photochemistry, daguerreotypes, evolution of cameras, film, more. Also artistic aspects—landscape, portraits, fine art, etc. 281 black-and-white photographs. 26 in color. 176pp. 8⅜ × 11¼. 25128-4 Pa. $13.95

THE DORÉ BIBLE ILLUSTRATIONS, Gustave Doré. 241 detailed plates from the Bible: the Creation scenes, Adam and Eve, Flood, Babylon, battle sequences, life of Jesus, etc. Each plate is accompanied by the verses from the King James version of the Bible. 241pp. 9 × 12. 23004-X Pa. $9.95

WANDERINGS IN WEST AFRICA, Richard F. Burton. Great Victorian scholar/adventurer's invaluable descriptions of African tribal rituals, fetishism, culture, art, much more. Fascinating 19th-century account. 624pp. 5⅜ × 8½. 26890-X Pa. $12.95

FLATLAND, E. A. Abbott. Intriguing and enormously popular science-fiction classic explores the complexities of trying to survive as a two-dimensional being in a three-dimensional world. Amusingly illustrated by the author. 16 illustrations. 103pp. 5⅜ × 8½. 20001-9 Pa. $2.50

THE HISTORY OF THE LEWIS AND CLARK EXPEDITION, Meriwether Lewis and William Clark, edited by Elliott Coues. Classic edition of Lewis and Clark's day-by-day journals that later became the basis for U.S. claims to Oregon and the West. Accurate and invaluable geographical, botanical, biological, meteorological and anthropological material. Total of 1,508pp. 5⅜ × 8½.
21268-8, 21269-6, 21270-X Pa. Three-vol. set $26.85

LANGUAGE, TRUTH AND LOGIC, Alfred J. Ayer. Famous, clear introduction to Vienna, Cambridge schools of Logical Positivism. Role of philosophy, elimination of metaphysics, nature of analysis, etc. 160pp. 5⅜ × 8½. (Available in U.S. and Canada only) 20010-8 Pa. $3.95

MATHEMATICS FOR THE NONMATHEMATICIAN, Morris Kline. Detailed, college-level treatment of mathematics in cultural and historical context, with numerous exercises. For liberal arts students. Preface. Recommended Reading Lists. Tables. Index. Numerous black-and-white figures. xvi + 641pp. 5⅜ × 8½.
24823-2 Pa. $11.95

HANDBOOK OF PICTORIAL SYMBOLS, Rudolph Modley. 3,250 signs and symbols, many systems in full; official or heavy commercial use. Arranged by subject. Most in Pictorial Archive series. 143pp. 8⅛ × 11. 23357-X Pa. $6.95

INCIDENTS OF TRAVEL IN YUCATAN, John L. Stephens. Classic (1843) exploration of jungles of Yucatan, looking for evidences of Maya civilization. Travel adventures, Mexican and Indian culture, etc. Total of 669pp. 5⅜ × 8½.
20926-1, 20927-X Pa., Two-vol. set $11.90

DEGAS: An Intimate Portrait, Ambroise Vollard. Charming, anecdotal memoir by famous art dealer of one of the greatest 19th-century French painters. 14 black-and-white illustrations. Introduction by Harold L. Van Doren. 96pp. 5⅜ × 8½.
25131-4 Pa. $4.95

PERSONAL NARRATIVE OF A PILGRIMAGE TO ALMANDINAH AND MECCAH, Richard Burton. Great travel classic by remarkably colorful personality. Burton, disguised as a Moroccan, visited sacred shrines of Islam, narrowly escaping death. 47 illustrations. 959pp. 5⅜ × 8½. 21217-3, 21218-1 Pa., Two-vol. set $19.90

PHRASE AND WORD ORIGINS, A. H. Holt. Entertaining, reliable, modern study of more than 1,200 colorful words, phrases, origins and histories. Much unexpected information. 254pp. 5⅜ × 8½. 20758-7 Pa. $5.95

THE RED THUMB MARK, R. Austin Freeman. In this first Dr. Thorndyke case, the great scientific detective draws fascinating conclusions from the nature of a single fingerprint. Exciting story, authentic science. 320pp. 5⅜ × 8½. (Available in U.S. only) 25210-8 Pa. $6.95

AN EGYPTIAN HIEROGLYPHIC DICTIONARY, E. A. Wallis Budge. Monumental work containing about 25,000 words or terms that occur in texts ranging from 3000 B.C. to 600 A.D. Each entry consists of a transliteration of the word, the word in hieroglyphs, and the meaning in English. 1,314pp. 6⅛ × 10.
23615-3, 23616-1 Pa., Two-vol. set $35.90

THE COMPLEAT STRATEGYST: Being a Primer on the Theory of Games of Strategy, J. D. Williams. Highly entertaining classic describes, with many illustrated examples, how to select best strategies in conflict situations. Prefaces. Appendices. xvi + 268pp. 5⅜ × 8½. 25101-2 Pa. $6.95

THE ROAD TO OZ, L. Frank Baum. Dorothy meets the Shaggy Man, little Button-Bright and the Rainbow's beautiful daughter in this delightful trip to the magical Land of Oz. 272pp. 5⅜ × 8. 25208-6 Pa. $5.95

POINT AND LINE TO PLANE, Wassily Kandinsky. Seminal exposition of role of point, line, other elements in non-objective painting. Essential to understanding 20th-century art. 127 illustrations. 192pp. 6½ × 9¼. 23808-3 Pa. $5.95

LADY ANNA, Anthony Trollope. Moving chronicle of Countess Lovel's bitter struggle to win for herself and daughter Anna their rightful rank and fortune—perhaps at cost of sanity itself. 384pp. 5⅜ × 8½. 24669-8 Pa. $8.95

EGYPTIAN MAGIC, E. A. Wallis Budge. Sums up all that is known about magic in Ancient Egypt: the role of magic in controlling the gods, powerful amulets that warded off evil spirits, scarabs of immortality, use of wax images, formulas and spells, the secret name, much more. 253pp. 5⅜ × 8½. 22681-6 Pa. $4.50

THE DANCE OF SIVA, Ananda Coomaraswamy. Preeminent authority unfolds the vast metaphysic of India: the revelation of her art, conception of the universe, social organization, etc. 27 reproductions of art masterpieces. 192pp. 5⅜ × 8½.
24817-8 Pa. $5.95

THE ART NOUVEAU STYLE BOOK OF ALPHONSE MUCHA: All 72 Plates from "Documents Decoratifs" in Original Color, Alphonse Mucha. Rare copyright-free design portfolio by high priest of Art Nouveau. Jewelry, wallpaper, stained glass, furniture, figure studies, plant and animal motifs, etc. Only complete one-volume edition. 80pp. 9⅜ × 12¼. 24044-4 Pa. $9.95

ANIMALS: 1,419 COPYRIGHT-FREE ILLUSTRATIONS OF MAMMALS, BIRDS, FISH, INSECTS, ETC., edited by Jim Harter. Clear wood engravings present, in extremely lifelike poses, over 1,000 species of animals. One of the most extensive pictorial sourcebooks of its kind. Captions. Index. 284pp. 9 × 12. 23766-4 Pa. $9.95

OBELISTS FLY HIGH, C. Daly King. Masterpiece of American detective fiction, long out of print, involves murder on a 1935 transcontinental flight—"a very thrilling story"—NY Times. Unabridged and unaltered republication of the edition published by William Collins Sons & Co. Ltd., London, 1935. 288pp. 5⅜ × 8½. (Available in U.S. only) 25036-9 Pa. $5.95

VICTORIAN AND EDWARDIAN FASHION: A Photographic Survey, Alison Gernsheim. First fashion history completely illustrated by contemporary photographs. Full text plus 235 photos, 1840–1914, in which many celebrities appear. 240pp. 6½ × 9¼. 24205-6 Pa. $8.95

THE ART OF THE FRENCH ILLUSTRATED BOOK, 1700–1914, Gordon N. Ray. Over 630 superb book illustrations by Fragonard, Delacroix, Daumier, Doré, Grandville, Manet, Mucha, Steinlen, Toulouse-Lautrec and many others. Preface. Introduction. 633 halftones. Indices of artists, authors & titles, binders and provenances. Appendices. Bibliography. 608pp. 8⅛ × 11¼. 25086-5 Pa. $24.95

THE WONDERFUL WIZARD OF OZ, L. Frank Baum. Facsimile in full color of America's finest children's classic. 143 illustrations by W. W. Denslow. 267pp. 5⅜ × 8½. 20691-2 Pa. $7.95

FOLLOWING THE EQUATOR: A Journey Around the World, Mark Twain. Great writer's 1897 account of circumnavigating the globe by steamship. Ironic humor, keen observations, vivid and fascinating descriptions of exotic places. 197 illustrations. 720pp. 5⅜ × 8½. 26113-1 Pa. $15.95

THE FRIENDLY STARS, Martha Evans Martin & Donald Howard Menzel. Classic text marshalls the stars together in an engaging, non-technical survey, presenting them as sources of beauty in night sky. 23 illustrations. Foreword. 2 star charts. Index. 147pp. 5⅜ × 8½. 21099-5 Pa. $3.95

FADS AND FALLACIES IN THE NAME OF SCIENCE, Martin Gardner. Fair, witty appraisal of cranks, quacks, and quackeries of science and pseudoscience: hollow earth, Velikovsky, orgone energy, Dianetics, flying saucers, Bridey Murphy, food and medical fads, etc. Revised, expanded In the Name of Science. "A very able and even-tempered presentation."—The New Yorker. 363pp. 5⅜ × 8. 20394-8 Pa. $6.95

ANCIENT EGYPT: ITS CULTURE AND HISTORY, J. E Manchip White. From pre-dynastics through Ptolemies: society, history, political structure, religion, daily life, literature, cultural heritage. 48 plates. 217pp. 5⅜ × 8½. 22548-8 Pa. $5.95

SIR HARRY HOTSPUR OF HUMBLETHWAITE, Anthony Trollope. Incisive, unconventional psychological study of a conflict between a wealthy baronet, his idealistic daughter, and their scapegrace cousin. The 1870 novel in its first inexpensive edition in years. 250pp. 5⅜ × 8½. 24953-0 Pa. $6.95

LASERS AND HOLOGRAPHY, Winston E. Kock. Sound introduction to burgeoning field, expanded (1981) for second edition. Wave patterns, coherence, lasers, diffraction, zone plates, properties of holograms, recent advances. 84 illustrations. 160pp. 5⅜ × 8¼. (Except in United Kingdom) 24041-X Pa. $3.95

INTRODUCTION TO ARTIFICIAL INTELLIGENCE: SECOND, ENLARGED EDITION, Philip C. Jackson, Jr. Comprehensive survey of artificial intelligence—the study of how machines (computers) can be made to act intelligently. Includes introductory and advanced material. Extensive notes updating the main text. 132 black-and-white illustrations. 512pp. 5⅜ × 8½. 24864-X Pa. $8.95

HISTORY OF INDIAN AND INDONESIAN ART, Ananda K. Coomaraswamy. Over 400 illustrations illuminate classic study of Indian art from earliest Harappa finds to early 20th century. Provides philosophical, religious and social insights. 304pp. 6⅝ × 9⅝. 25005-9 Pa. $11.95

THE GOLEM, Gustav Meyrink. Most famous supernatural novel in modern European literature, set in Ghetto of Old Prague around 1890. Compelling story of mystical experiences, strange transformations, profound terror. 13 black-and-white illustrations. 224pp. 5⅜ × 8½. (Available in U.S. only) 25025-3 Pa. $6.95

PICTORIAL ENCYCLOPEDIA OF HISTORIC ARCHITECTURAL PLANS, DETAILS AND ELEMENTS: With 1,880 Line Drawings of Arches, Domes, Doorways, Facades, Gables, Windows, etc., John Theodore Haneman. Sourcebook of inspiration for architects, designers, others. Bibliography. Captions. 141pp. 9 × 12. 24605-1 Pa. $7.95

BENCHLEY LOST AND FOUND, Robert Benchley. Finest humor from early 30's, about pet peeves, child psychologists, post office and others. Mostly unavailable elsewhere. 73 illustrations by Peter Arno and others. 183pp. 5⅜ × 8½. 22410-4 Pa. $4.95

ERTÉ GRAPHICS, Erté. Collection of striking color graphics: *Seasons, Alphabet, Numerals, Aces* and *Precious Stones*. 50 plates, including 4 on covers. 48pp. 9⅜ × 12¼. 23580-7 Pa. $7.95

THE JOURNAL OF HENRY D. THOREAU, edited by Bradford Torrey, F. H. Allen. Complete reprinting of 14 volumes, 1837–61, over two million words; the sourcebooks for *Walden*, etc. Definitive. All original sketches, plus 75 photographs. 1,804pp. 8½ × 12¼. 20312-3, 20313-1 Cloth., Two-vol. set $125.00

CASTLES: THEIR CONSTRUCTION AND HISTORY, Sidney Toy. Traces castle development from ancient roots. Nearly 200 photographs and drawings illustrate moats, keeps, baileys, many other features. Caernarvon, Dover Castles, Hadrian's Wall, Tower of London, dozens more. 256pp. 5⅜ × 8¼. 24898-4 Pa. $6.95